Preface

Electrochemical power sources range from miniature batteries with an energy storage capability of less than 0.1 Wh to projected load levelling modules with capacities greater than 10 MWh. All such systems utilize the energy evolved by spontaneous chemical reactions to produce electric power **directly**, and all rely on the same fundamental physical processes for their operation. Commercial batteries have now been manufactured for over a century, but until comparatively recently, research and development in the battery industry has been directed largely towards improvements in well tried systems, especially in the fields of engineering design and production. However recent advances in electrochemistry and materials science have opened the way for the evolution of entirely new types of power source with greatly improved electrical performance and other desirable characteristics.

Our aim in writing this book has been to produce a brief but comprehensive account which may be read – we hope with some profit and interest – by anyone with a basic knowledge of chemistry and physics who wishes to know something about how batteries work, and what the main developments are in this fascinating area of science and technology. We have tried to maintain a balance between describing well established 'conventional' systems, and 'state-of-the-art' developments which may or may not become of commercial importance. However because of the existence of excellent specialised texts which describe in some detail the evolution of the main commercial batteries, we have placed considerable emphasis on discussing recent trends and discoveries.

As the book has been written for the non-specialist, the theoretical background to the basic processes involved in cell operation is described in some detail in preference to a more thorough series of comparisons of the characteristics and performance of competing systems. We have excluded any discussion on the very closely related field of fuel cells since a number of accounts of this topic have been published recently. It has been our intention to describe and characterise most of the established and emerging primary and secondary battery systems which are of current commercial or theoretical interest. Research into novel power sources may shortly lead to the major breakthroughs necessary before electric vehicles become a major component of the transportation system, and electrochemical storage becomes a serious competitor for large scale load levelling applications. It is our hope that this book may lead to a more general appreciation of the rôle of the electrochemical power source today and its potentialities in the future.

We have to thank our many friends both in industry and in research laboratories for their assistance in the preparation of this book. In particular we would like to acknowledge Dr J. Thompson (Royal Aircraft Establishment (MOD), Farnborough) who reviewed the whole manuscript, and the following colleagues who gave us valuable advice: Dr P. Bruni (Superpila), Dr W.G. Bugden (British Rail), Dr Ing. G. Clerici (Magneti Marelli), Dr M.D. Ingram (University of Aberdeen), Dr R. Marassi (University of Camerino), Dr B.B. Owens (Medtronic) and Dr B. Rivolta (Polytecnic of Milan). We would also wish to thank the battery companies from all corners of the globe who were so generous in their provision of technical information, photographs and diagrams. Finally we thank Ann Hughes of the University of St. Andrews for her patience in typing and re-typing the manuscript.

It may be considered appropriate to mention the Scottish/Italian collaboration which has resulted in the writing of this book. It started in 1965 when two of us (BS and CAV) were working in the laboratories of H.A. Laitinen at the University of Illinois and expanded to include members of the CNR Centre on Electrode Processes at Milan when a joint research programme was instituted some ten years ago. We are pleased to acknowledge the financial assistance given by the Science and Engineering Research Council (UK), the Consiglio Nazionale delle Ricerche (Italy), NATO and the British Council to foster this collaborative venture over the years.

St. Andrews
June 1982

urces

Edward Arnold

First published in Great Britain 1984
by Edward Arnold (Publishers) Ltd
41 Bedford Square
London WC1 3DQ

Edward Arnold
300 North Charles Street
Baltimore
Maryland 21201
USA

Edward Arnold (Australia) Pty Ltd
80 Waverly Road
Caulfield East 3145
PO Box 234
Melbourne

British Library Cataloguing in Publication Data

Modern batteries.
   1. Electric batteries
   I. Vincent, C.A.
   621.31'242   TK2901

   ISBN 0–7131–3469–0

Text set in 10/11pt Times Roman
by Castlefield Press, Moulton, Northampton
Printed and Bound by
Thomson Litho Ltd, East Kilbride, Scotland

Contents

1 Introduction

1.1 Electrochemical power sources

An **electrochemical power source** or **battery** is a device which enables the energy liberated in a chemical reaction to be converted directly into electricity. Batteries fulfil two main functions. First and foremost they act as portable sources of electric power. Well known modern examples range from the small **button cells** used in electric watches to the **lead–acid batteries** used for starting, lighting and ignition in vehicles with internal combustion engines. The second function, which is likely to increase in importance over the next twenty years, is based on the ability of certain electrochemical systems to store electrical energy supplied by an external source. Such batteries may be used for driving electric vehicles, for emergency power supplies, and as part of the main electricity supply system for meeting short duration demand peaks (**load levelling**) or in conjunction with **renewable energy sources**, such as solar, wave or wind power.

The first authenticated description of an electrochemical battery was given by Alessandro Volta, Professor of Natural Philosophy (Physics) at the University of Pavia in Italy, in a letter to the Royal Society (London) in 1800. A photograph of an original Volta 'pile' is shown in Fig. 1.1. The importance of Volta's discovery as a tool for advancing the understanding of chemistry and physics was immediately grasped by scientists in a number of countries. However it was the introduction of telegraph systems, which were becoming of increasing importance in the 1830's, that gave rise to the development of reliable commercial batteries, capable of sustaining a substantial flow of current, without undue loss of cell voltage.

The first high current electroplating battery was described in 1840 and over the next two decades the use of techniques such as electroplating and electroforming, together with the exploitation of practical electric motors gradually became more widespread. In the 1870's a more general consumer market for batteries was created by the manufacture of electric bell circuits for homes, offices and hotels. The 'flash light' was introduced at the turn of the century, some twenty years after Edison's invention of the incandescent lamp. By then the annual production of batteries in the USA alone had exceeded two million units.

The large scale introduction from 1870 onwards of dynamos or electro-magnetic generators driven by heat engines led to the worldwide industrial and domestic use of 'mains' electricity, accepted as common-place today, a century later. This ready availability of electrical energy was the principal

1.1   Original Volta pile (By courtesy of the Tempio Voltiano, Como, Italy)

motivation for the development of secondary or storage batteries, although this area was soon to be further stimulated by the demands of the growing motor car industry.

A further impetus for commercial battery development came with the introduction of domestic radio receivers in the 1920's and an equivalent growth has been seen over the last twenty years with the development of solid state microelectronic equipment. Today it is estimated that annual battery production totals 8-15 units per head of population throughout the developed countries of the world.

1.2 Nomenclature

There is some confusion in the terminology used to denote the electro-chemical devices which convert chemical into electrical energy. In many cases the devices have changed in character with the passage of time, but have retained their original names; in others, the terms commonly used do not clearly define the nature of the device. In this book we conform to general usage and employ the words **cell** and **battery** interchangeably to describe a **closed electrochemical power source** – i.e. one in which the reactants are incorporated during manufacture. The term 'battery' originally implied a group of 'cells' in a series or parallel arrangement, but is now understood to mean either a single cell or a group of cells.

Two other terms which are not self-explanatory are **primary** and **secondary** cell. A primary system is one whose useful life is ended once its reactants have been consumed by the discharge process. In contrast, a secondary system is capable of being **charged** or **recharged** when its reactants have been used up: the spontaneous electrochemical reaction can be reversed by passing current through the cell in the opposite direction to that of cell discharge. A secondary battery might therefore be considered as an electrochemical energy storage unit. Note, however, that the energy derived from the external current is stored as chemical energy, and not as electrical energy as in a capacitor. Other terms are sometimes used to describe this type of system – e.g. 'accumulator' (introduced by Davy along with the terms 'cell' and 'circuit'), 'storage battery', 'rechargeable battery', etc.

We do not consider the related subject of **fuel cells** where both cathodic and anodic reagents – usually gases – are stored externally, and can be supplied to the electrochemical cell on a continuous basis. A number of books have recently been published on this topic. The term **hybrid cell** is used here to describe a power source in which one of the active reagents is in the gaseous state – e.g. the oxygen of the air. Use of the word 'hybrid' in this context should not be confused with its meaning in the phrase 'hybrid electric vehicle' which refers to an electric vehicle with more than one power supply, as described below.

A large number of technical terms are associated with the literature on batteries: the more common of these are given in the Glossary, while the electrical quantities used to describe battery performance and characteristics are defined in Section 2.5, and summarised in Appendix 2.

The most common convention for writing an electrochemical cell is to place the negative electrode on the left and the positive on the right. The cell is then named in the same way: thus reference to the 'sodium–sulphur cell' implies that sodium is the active reagent at the negative electrode and sulphur that at the positive electrode. We make three exceptions to this general rule in order to conform to normal usage, and call the lead–lead oxide cell, the 'lead–acid cell', the cadmium–nickel oxide cell the 'nickel–cadmium cell' and the zinc–manganese dioxide cell, the 'Leclanché cell'.

1.3 The renaissance in battery development

Until very recently, 'conventional' batteries using solid electrodes and aqueous electrolytes proved satisfactory for the majority of common applications. Traditional primary systems, such as the Leclanché $Zn–MnO_2$ cell and the Ruben–Mallory $Zn–HgO$ cell have been (and to a large extent still are) adequate power sources for most portable electrical equipment. Well established rechargeable batteries such as those based on the lead–acid or nickel–cadmium systems have for long been employed as small localised energy storage units (e.g. in rural areas, in railway and telephone systems, etc.) and as sources of auxiliary power in ground, air and sea transport. For many years research and development in the battery industry has been directed mainly towards improvements in these well known systems, especially in the fields of engineering design and production.

In the last fifteen years, however, the situation has changed considerably. First, advances in semiconductor technology have led to the production of large scale integrated (lsi) circuits in immense numbers, bringing about a revolution in the electronics industry. Microelectronic components are now inexpensive and are widely used in the production of pocket calculators, electric watches and similar devices. Japan alone produced 45 million electronic calculators in 1979; in the same year out of an estimated total watch sale of 2.6×10^8 worldwide, over one third were electrically driven. Development of such miniature electronic instruments demanded the evolution of miniature power supplies which would offer a much higher energy per unit volume and superior discharge characteristics as compared with those of traditional batteries.

The second, and perhaps more important factor affecting the demand for new battery systems was the realisation in the late 1960's that the constantly increasing energy needs of the developed countries of the world would soon lead to the progressive exhaustion of oil supplies. This in turn led to the requirement for more efficient use of the remaining fossil fuel reserves and for a shift towards the exploitation of alternative energy sources, preferably of a clean and regenerative type. Central to the problem both of the utilisation of discontinuous energy sources – for example solar, wind and wave power, and of the efficient running of conventional generating plant is the provision of suitable energy storage systems. While there are a number of alternatives to be considered such as pumped hydroelectric or compressed-air storage, electrochemical storage batteries are in many instances more convenient, being transportable and flexible in size, as well as being silent and non-polluting. For this application, batteries require the ability to undergo large numbers of deep charge/discharge cycles with high efficiency and without physical degradation.

Since a considerable proportion of all petroleum is consumed rather inefficiently in vehicle traction, the possibility of replacing vehicles driven by internal combustion engines by battery powered electric transport is under active consideration, and development of advanced batteries for this purpose is being pursued in a number of countries. Since batteries for electric vehicles (EVs) must be transported as part of the vehicle load, they require high power/mass ratios in addition to high cycle efficiency.

1.4   A survey of common battery types and applications

The **total available energy** of a battery is a measure of how much electricity it can deliver (usually measured in Wh) and is directly related to the size of the unit. One possible classification of the most common commercial battery types, according to size, is given in Table 1.1. The range of battery energies extends over at least fifteen orders of magnitude. At the bottom of the range there are 0.1 cm^2 experimental cells with a PbF_2 solid electrolyte which have a total energy of just over 1×10^{-6} Wh. The smallest commercial button cells have energies of around 100 mWh, while the common 'D-size' cylindrical cells which have a total volume of 45 cm^3 have energies in the range 2-15 Wh.

Table 1.1
Classification of batteries according to size

Type	Energy	Applications
Miniature batteries	100 mWh–2 Wh	Electric watches, calculators, implanted medical devices
Batteries for portable equipment	2 Wh–100 Wh	Flash-lights, toys, power tools, portable radio and television
SLI batteries (starting, lighting and ignition)	100–600 Wh	Cars, trucks, buses, tractors. Lawn mower traction
Vehicle traction batteries	20–630 kWh (3 MWh)	Fork-lift trucks, milk-floats, locomotives (Submarines)
Stationary batteries	250 Wh–5 MWh	Emergency power supplies, local energy storage, remote relay stations
Load levelling batteries	5–100 MWh	Spinning reserve, peak shaving, load levelling

Rechargeable cells used in power tools and other 'cordless' electric appliances can supply 20-100 Wh. At the top of the range, submarine lead–acid batteries weighing nearly 200 tonnes have rated energies of 3 MWh while the specification for projected load levelling batteries is 100 MWh.

Miniature batteries

Miniature batteries based on aqueous, non-aqueous and solid electrolytes are manufactured as power sources for microelectronic and other mini-aturised equipment. In Fig. 1.2 the sizes and shapes of some representative button cells are shown. A typical application for such cells is in the electronic watch, where the oscillator circuit draws a continuous current of $0.2–0.6\,\mu A$ and depending on the type of frequency divider and display, the complete unit may require a total of up to $0.5–2.0\,\mu A$ for operation. Hence the total amount of electrical energy consumed in driving the watch for a year is in the range 15–60 mWh. At present batteries are manufactured which last for 5-7 years; the ten year watch battery will soon be produced. Watch batteries must have exceptionally low self-discharge rates and very reliable seals to prevent leakage. Further, they have strict design constraints, in order to fit into the restricted available space within the watch case. From an electrical point of view, as well as having a high volumetric energy density, (i.e. amount of energy delivered per unit volume) the battery must display a flat

30-50 mAh

| 2.0 x 9.5 | 2.0 x 11.6 | 2.0 x 12.5 | 2.0 x 16.0 | 1.6 x 20.0 |

100-170 mAh

| 1.6 x 20.0 | 1.8 x 23.0 | 1.9 x 20.0 | 2.0 x 23.0 | 2.0 x 24.5 |

| 2.5 x 20.0 | 3.0 x 24.5 | 3.2 x 20 |

200-250 mAh   **450 mAh**   **850 mAh**

| 2.0 x 27.0 | 3.0 x 24.5 | 3.81 x 27.0 | 3.96 x 37.0 |

1.2 Dimensions (in mm) and capacities of some representative button cells.

discharge characteristic for accurate operation of the watch circuits. In addition, many watch batteries must be able to sustain occasional high rate discharge pulses, since watches with liquid crystal diode displays usually have a small tungsten lamp for back lighting. An alternative approach for the provision of power for watches is to use a miniature battery of relatively small capacity, but which can be recharged using solar cells.

An excellent example of the effects of the simultaneous developments of microelectronic circuitry and battery technology has been the increase in service life and reduction in physical size of cardiac pacemakers. In 1979 there were more than 250 000 of these devices in use. The earliest pacemakers, implanted in the 1960s, were simple pulse generators operating at fixed rates, for periods of 1-3 years. They were powered by zinc–mercury oxide batteries, and the whole unit weighed about 200 g. Over the last decade there has been a major switch to batteries based on lithium anodes with a much higher practical energy density than conventional aqueous systems. This, together with further miniaturisation of the electronic components, has led to a reduction in mass of the complete unit to 50 g (25 g units are now planned), with a projected lifetime of almost ten years. At the same time, the sophistication of the electronic circuits has greatly increased, and 'demand' pacemakers, whose pulsing is inhibited by physiological signals indicating normal ventricular activity, are now the norm. Appliances under development include units which are programmable from outside the body for sensitivity, pulse rate, width and amplitude, etc., and others which rely on a more extensive monitoring of several physiological signals and which are able to select appropriate stimulation strategies in response to the input. Yet another related development is that of implantable automatic

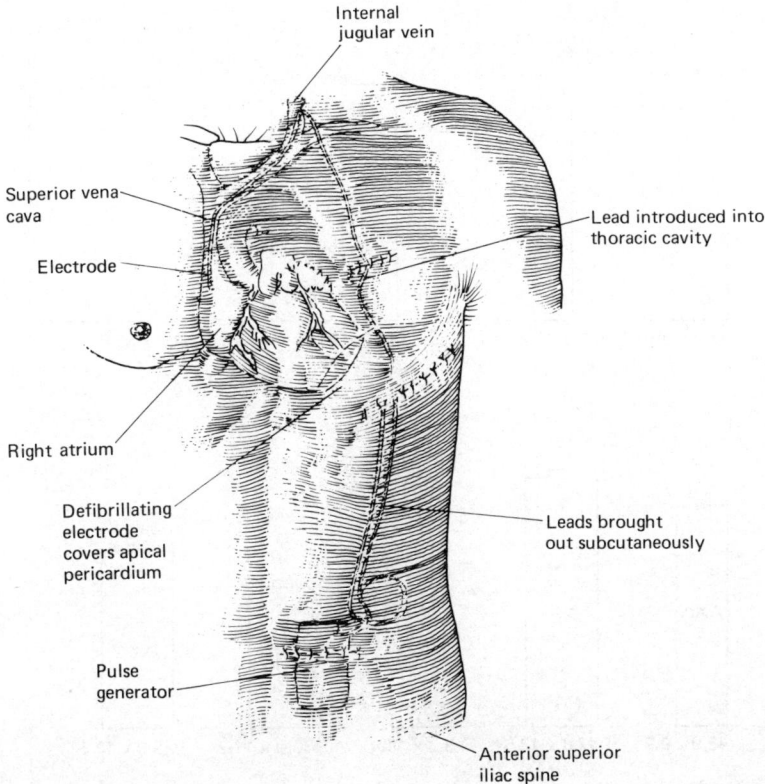

1.3   Schematic illustration of the implantation of an automatic cardiac defibrillator. (By permission of World Medicine)

antifibrillation devices, (Fig. 1.3). By electrically monitoring the activity of the heart, the unit can detect the onset of ventricular fibrillation. It then applies a large defibrillating current pulse which shocks the heart into normal rhythm. The power levels of present day demand pacemakers are quite low: 25-50 μW for sensing and 60-100 μW for stimulation. Much higher pulse power levels will be required for next generation devices such as the defibrillators, so that higher rate batteries must be developed.

Batteries for portable equipment

Perhaps the best known of all batteries are the inexpensive packaged 'dry batteries' used in a wide variety of portable lighting appliances, battery operated toys, transistorised radio receivers and numerous other applications. A number of standard sized cylindrical single cells are shown in Fig. 1.4. The majority of this class of battery are based on the Leclanché Zn–MnO_2 system. Although the original was introduced towards the end of the last century, there have been (and still are) notable and continuing

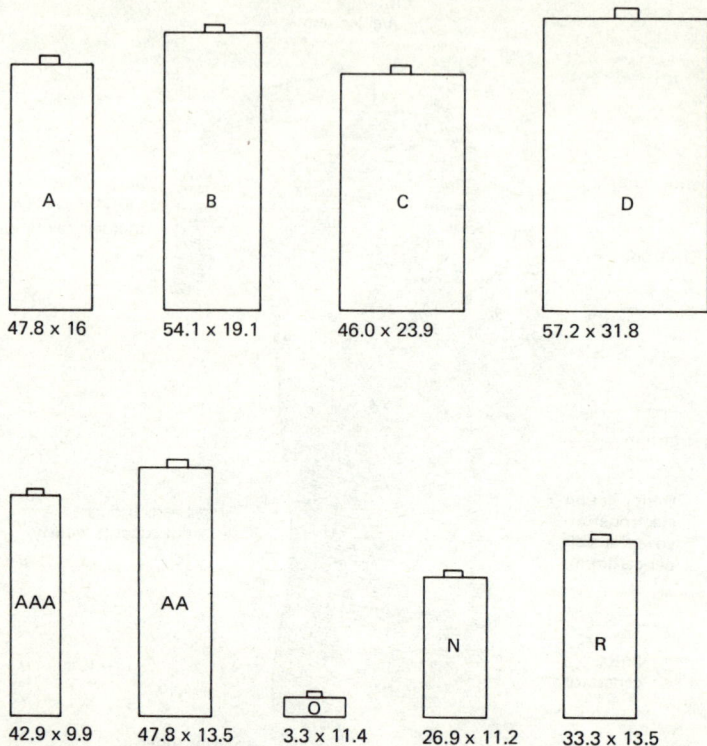

A 47.8 x 16
B 54.1 x 19.1
C 46.0 x 23.9
D 57.2 x 31.8

AAA 42.9 x 9.9
AA 47.8 x 13.5
O 3.3 x 11.4
N 26.9 x 11.2
R 33.3 x 13.5

1.4 Standard dimensions (in mm) for cylindrical primary cells.

improvements in the operating characteristics of this system. However such batteries tend to have rather poor voltage regulation; i.e. the cell voltage decreases as discharge through a fixed load progresses. The introduction of the Ruben–Mallory Zn–HgO cells in the 1940's provided a power source with a very flat discharge characteristic and an improved energy density. More recently, even higher energy densities have been achieved by cells having lithium anodes and electrolytes based on organic solvents. Such batteries are relatively expensive compared with Leclanché batteries, but are used extensively where light weight is important – e.g. for portable radio transceivers and tape-recorders, night vision equipment, missile and other rocket-borne systems, etc.

For many applications where a relatively high current drain is involved, it is much more cost effective to use a sealed or 'maintenance free' rechargeable battery as an electrical energy store. This is true, for example, for 'cordless' power tools, portable television receivers, hedge trimmers and lawn edging tools, emergency lights, etc. Examples of such batteries are the alkaline nickel–cadmium, the alkaline iron–nickel oxide or the sealed lead–acid systems. Many of these batteries in the 10 mAh–15 Ah range have exactly the same external dimensions as the primary cells they are designed to replace. Larger units with capacities up to 1000 Ah are also available. The alkaline batteries have an excellent cycle life, (i.e. they can undergo a large number of charge/discharge cycles) perform well at very low temperatures, have low self-discharge rates, good voltage regulation and are very robust. They are however much more expensive than their lead–acid equivalents. Both varieties suffer from poor energy density (i.e. amount of energy delivered per unit mass) and may be replaced in the future by lithium–organic systems for applications where light weight is desirable.

SLI batteries

Almost a hundred million batteries for starting, lighting and ignition (SLI) in internal combustion engined vehicles are manufactured annually, using over one third of the total world output of lead. Such batteries are required to 'turn' engines with high compression, often at low temperatures where the viscosity of engine oil is high: currents of up to 200 A may be drawn for this purpose. In addition the battery must supply power for ignition, lighting, ventilation, rear window heaters, etc. Most modern car electric circuits employ a nominal 12 V system and the batteries have six lead–acid cells in series, with capacities of the order of 100 Ah. Significant improvements in materials and design have raised the energy density of SLI batteries to around 45 Wh kg^{-1} and 75 Wh dm^{-3}, while retaining durability and service life. (It is inappropriate to talk of 'cycle life' in this context since it is unusual for SLI batteries to undergo deep discharge). Work is continuing in a number of centres on the development of improved 'maintenance-free' SLI batteries which require attention only once every 3-5 years. Because of their large scale of manufacture, SLI batteries are relatively inexpensive and are therefore often used quite successfully for other applications, e.g. to provide tractive power for lawn mowers and 'go-karts', in emergency lighting units, etc.

Nickel–cadmium batteries with thin sintered plates are used for on-board power supplies in aircraft, helicopters, tanks and military vehicles where their excellent low temperature high rate performance is an important attribute. Modern 40 Ah cells designed for airborne use can deliver 20 kW of instantaneous power at 25°C and over 10 kW at −30°C. Again, the high cost of this system compared with that of lead–acid batteries has restricted its use.

Vehicle traction batteries

As pointed out above, the advantages of widespread use of electric batteries for vehicle traction are numerous: in addition to improvements in the environment effected by the silent, pollution-free operation of EVs, conservation of oil supplies and a more efficient performance of the mains electricity generating system, brought about by load levelling due to overnight charging of EV batteries, would result. In the United Kingdom there are more than 40 000 electric road vehicles – most of them 'milk floats' (bottled-milk delivery trucks). In addition there is a wide variety of fork-lift trucks, airport tractors, warehouse and baggage trucks, etc. However there are only a few, mainly experimental urban delivery EVs and although many electric cars have been designed, there is currently no major production line in operation. It may be of interest to note than an electric car held the world land speed record at 66 mph (106 km h^{-1}) in 1899; this was increased to 85 mph (136 km h^{-1}) two years later. Present day cars have reached 150 mph (240 km h^{-1}). EVs have fewer moving parts than other vehicles and the inherent lack of vibration of the electric motor cuts down wear, reducing the need for maintenance and extending the vehicle's life. Electronic control and regenerative braking made EVs easy to drive, especially in heavy urban traffic. The reason for the relative lack of commercial success of EVs up to the present lies in the battery specifications required in order to produce a performance (and cost) even approaching those of internal combustion powered vehicles. The speed, acceleration and range of most present day EVs are limited by the low energy and power densities of lead–acid traction batteries. Such batteries have an even poorer energy density than SLI batteries (around 20-30 Wh kg^{-1}) but their tubular electrodes permit extended deep cycling.

A typical family car, having a laden weight of 1-1.5 tonnes would require 5-10 kWh of energy to move 50 km. Taking the lower figure, this would involve the consumption of approximately 4.5 dm^3 or 3.9 kg of petrol. An average value for the energy density of a lead–acid battery capable of deep cycling at a useful rate might be 25 Wh kg^{-1}. Hence a battery weighing some 200 kg and occupying 120 dm^3 would be required to drive the car the same distance. In other words the lead–acid energy store is up to fifty times heavier and uses up twenty-five times the space of its petrol equivalent. Other disadvantages of EVs include the restricted power available for acceleration and hill-climbing and the time taken to recharge the battery. One might compare this latter period of 6-12 hours with the 2-3 minutes needed to refill a petrol tank: it is interesting to calculate that the equivalent power flow through a normal petrol pump delivery hose is about 30 MW!

Recently prototype lead–acid traction batteries with energy densities in the range 40-60 Wh kg^{-1} have been announced by several manufacturers; however it seems unlikely that batteries of this type which are capable of more than 1000 deep cycles will ever have energy densities much above 40 Wh kg^{-1}. On the other hand, a number of advanced batteries are under development for EV use: two are ambient temperature batteries, zinc–nickel oxide (75 Wh kg^{-1}) and zinc–chlorine hydrate (80 Wh kg^{-1}), and two are high temperature systems, sodium–sulphur (120 Wh kg^{-1}) and lithium–iron sulphide (100 Wh kg^{-1}). In the longer term one or more of these systems is likely to give three to four times the range of a lead–acid EV, at lower initial cost, higher efficiency and, it is hoped, equal cycle life. In addition to developments in battery technology the future of the electric road vehicle is dependent on a wide range of other variables including the relative costs of electricity and petrol, availability of off-peak power, initial capital costs, pollution controls, etc. In the short term, the most realistic target for a commercially competitive EV is likely to be an urban delivery van with a payload of 2 tonnes and a range of about 120-150 km.

A **hybrid electric vehicle** has more than one type of power supply to support the drive: these may be used independently, in series or in parallel to increase the range of the vehicle or to improve its overall efficiency – for example by recovering energy by means of regenerative braking. The combinations of battery plus mains power and battery plus internal combustion engine have proved the most promising examples of practical hybrid EV power supplies, but the increased complexity and cost of these vehicles has prevented their commercial development so far. Battery/battery hybrids have also been tested: for example a high performance, but heavy, lead–acid battery may be used to provide power for acceleration and hill-climbing, while a light zinc–air battery with high energy density supplies the energy for running at constant speed.

Battery-powered trains or rail-cars have been in operation since the turn of the century, mainly in W. Germany. They have the advantage over electric road vehicles of much lower frictional losses, so that energy density limitations are not so severe. Modern rail-cars seating eighty passengers and operating on 630 kWh lead–acid batteries (which have a mass of 21 tonnes and a cycle life of 1250-1550), can accelerate at 0.4 m s^{-2} to 100 km h^{-1} on the level, and have a range of 250-450 km. High reliability, low maintenance costs and safety are the important attributes to this form of transport. The use of hybrid mains/battery rail cars is currently being investigated: for example, a relatively small 1 tonne sodium–sulphur battery would allow a rail-car to make a circuit of some 100 km away from an electrified line.

Submarines of the non-nuclear reactor type rely on storage batteries for propulsion in submerged condition. Generally the lead-acid system is used, but the light silver–zinc batteries have been installed in some vessels. The energy stored is in the range of 2-3 MWh, and the batteries can weigh more than 180 tonnes. Special precautions are required for such large units to ensure adequate heat dissipation and thus to prevent the build-up of excessive temperature differentials.

Stationary batteries

The main functions of storage batteries in the 250 Wh-5 MWh range are to provide stand-by power for emergency lighting, telephone services, hospital equipment, etc., or for local storage in remote relay stations, radio beacons, navigation buoys, etc. In both types of application it is rare for deep cycling to be involved. Generally a continuous 'trickle' charge is available, (see Appendix 1) either from mains power or from solar, wind or wave energy. Stationary batteries for stand-by power are usually constructed of a special design of lead–acid cell known as the 'high performance Planté' type. This type of battery generally has very low standing losses and very long lifetimes; 20-30 years are typical. About 25% of stand-by power is based on nickel–cadmium batteries.

A schematic circuit for an emergency power supply is shown in Fig. 1.5 The load is fed directly from the mains supply (which also 'float' charges the battery) but in the event of mains failure, the load is automatically switched to the inverter by the mains monitoring system.

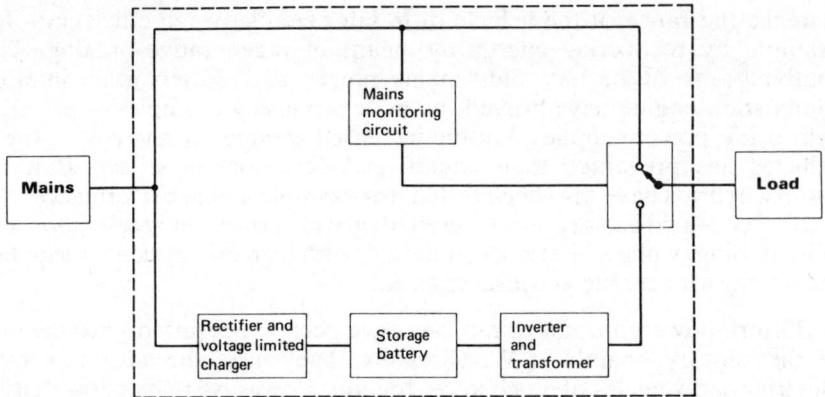

1.5   Schematic circuit for uninterruptible emergency power supply.

Windmill/battery supplies for automatic telephone exchanges have been successfully implemented in remote locations in Denmark using 400 Ah lead–acid batteries. Small wave-power air turbines ($\approx$100 W) have been used to charge batteries on maritime navigation buoys since 1965, and solar energy/battery systems have worked well in survey buoys and for supplying energy for light beacons near airports. Spacecraft batteries which are recharged by solar energy are shown in Fig. 4.25 and 4.29.

Batteries for load levelling

A schematic smoothed national electricity demand curve is shown in Fig. 1.6. Such a graph invariably shows a deep minimum during the night, followed by increased demand during waking hours. Super-imposed on this curve are a number of sharp spikes, usually associated with cooking and television watching habits. The electricity generating industry must therefore maintain plant capable of producing in excess of the peak power requirement. This is generally accomplished by generating:
  (i) the base load with nuclear reactors or other efficient generators running at their optimum capacity,
 (ii) the intermediate load, by adding older less efficient plant, normally at part load, and
(iii) peak demand, by bringing onto line expensive gas turbine-driven generators.
The principle of load levelling is to increase the base load capacity and to use the excess capacity available during the demand trough to stockpile energy which may be used to meet the peak demand later in the cycle. A number of techniques are currently employed to put load levelling into effect. In mountainous temperate countries pumped hydro-storage is economical despite high capital costs: individual schemes of 200-2000 MWh capacity have been running for many years, throughout the world. Other methods include thermal storage, compressed-air storage (in combination with a modified combustion turbine), hydrogen storage, etc.

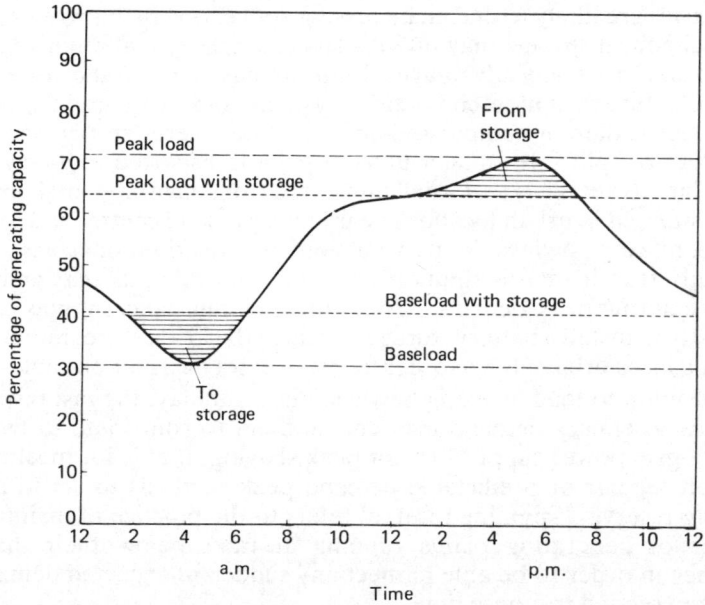

1.6 National electricity demand curve. In the winter the curve moves up the y-axis: in the United Kingdom the winter demand is roughly double that of summer.

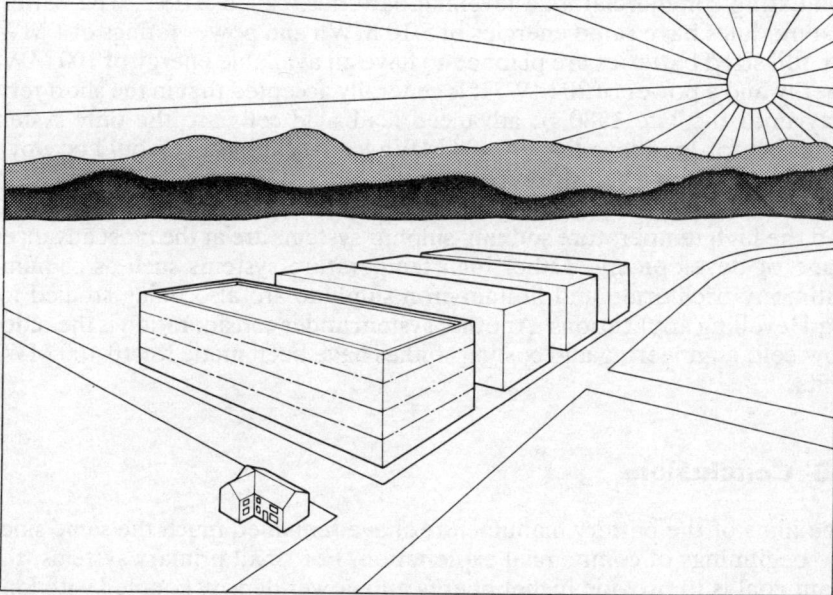

1.7 Artist's impression of a 100 MWh load levelling battery. (By courtesy of I.D.C. Cargill, University of St. Andrews) see also Fig. 6.27.

Batteries are likely to find an increasing application in this role. Although electrochemical storage may prove a less cost effective alternative in some situations, it has many advantages. Batteries have a much shorter lead time in manufacture than most competitive systems, and being modular (unlike a hydroelectric dam or compressed-air store), the energy storage facility can be added to, split into smaller units or even transported to a new site. A particular advantage is that small units, free from environmental problems, can be situated in urban locations near industrial load centres and so lead to very significant savings in transmission and distribution costs. This is especially true if the distribution network is reaching its maximum rated capacity during peak demand. Under these circumstances it may be much less costly to install a battery storage system at the load centre, rather than to disrupt a large urban area in order to install extra distribution cables.

In addition to load levelling between night and day, the fast response of batteries to energy demand may enable them to contribute to two other areas of grid power supply – (i) for peak shaving, that is for meeting short duration regular or predictable demand peaks and (ii) to act in place of spinning reserve. 'Spinning reserve' refers to the practice of maintaining a number of generating plants running at rates below their maximum efficiency in order to be able to meet any sudden unexpected demand, say with a ten second response time.

In the USA, the Electric Power Research Institute (EPRI) and the Energy Research and Development Administration (ERDA), now the US Department of Energy (DOE) have established a national test facility for evaluating commercial load levelling batteries, called BEST. The battery test modules have rated energies of 5-10 MWh and power ratings of 1 MW; the full sized batteries are planned to have an available energy of 100 MWh energy and a power of 20 MW. It is generally accepted that in the short term (say until the late 1980's), advanced lead–acid cells are the only system available for load levelling. A 100 MWh lead–acid battery would occupy a building two and a half storeys high and an area of about 0.25 km^2 (Fig. 1.7). Of the advanced load levelling batteries, the ambient zinc–chlorine hydrate and the high temperature sodium–sulphur systems are at the most advanced stage of development. Other high temperature systems such as sodium–antimony trichloride and lithium–iron sulphide are also being studied for load levelling applications. Another system under consideration is the redox flow cell: engineering and costing studies have been made for 10-100 MWh units.

1.5 Conclusions

The aims of the battery manufacturer have remained much the same since the beginnings of commercial exploitation. For small primary systems, the main goal is to provide higher energy and power density coupled with long shelf life and low cost. However primary batteries are always an expensive way of purchasing energy,* and further moves to replace many primary batteries with compatible rechargeable units are to be expected. For application in electric vehicles, energy density (again coupled with low cost) is the prime requirement. However for energy storage, cost alone must be the overriding consideration. The overall cost of energy storage can be divided into two components. First the initial capital cost of the battery must be considered, which includes raw materials, a contribution to research and development, investment in plant, administrative overheads and manufacturing costs. Provided that the battery has a lengthy cycle life, a high capital cost may be acceptable: a system capable of 1500 cycles and operated at 1 cycle per day, five times per week could be amortized over a 5-6 year period. The second component is determined by the efficiency of the charge/discharge cycle i.e. by the ratio of energy supplied by the battery during discharge to the energy required to charge it. The extensive exploitation of batteries in a load levelling role must await the development of a new long life, highly efficient system.

---

*In Scotland, the cost of domestic mains electricity is £0.0413 per kWh (in 1981); a D-size Leclanché cell, delivering say 5 Wh, currently retails at £0.24. Thus energy from the primary battery costs £48.00 per kWh – a factor of over 1000 more expensive. For a 150 mWh zinc–silver oxide button cell, retailing at £1.00, the cost of energy is ≈£7000 per kWh!

2 Theoretical Background

2.1 Introduction

Many chemical substances are able to exist in more than one oxidation state: that is to say, they can donate or accept electrons from other species. A net transfer of electrons such as

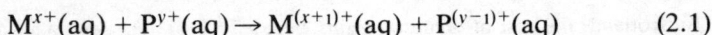

$$M^{x+}(aq) + P^{y+}(aq) \rightarrow M^{(x+1)+}(aq) + P^{(y-1)+}(aq) \qquad (2.1)$$

may occur spontaneously whenever the free energy of the whole system is reduced by such a process. An exchange of electrons can take place in homogeneous solution as above or, under suitable circumstances, electrons can be transferred to or from an electronic conductor. An electron transfer between an electronic conductor and some species in an electrolytic phase is known as an **electrochemical reaction** or **electrode process**. For such a transfer to proceed on a continuous basis, the principle of electroneutrality requires it to be associated with a second electrode process in which electron transfer takes place in the opposite direction. If the two compartments in the vessel shown in Fig. 2.1 contain solutions of M^{x+} (aq) and P^{y+} (aq) respectively, then connection of the two inert metal conductors through a resistor will permit charge flow round the circuit and hence formation of $M^{(x+1)+}(aq)$ and $P^{(y-1)+}(aq)$. As charge transfer (in the appropriate direction) at the interfaces produces exactly the same result as the homogeneous electron exchange, the whole **cell process** including the electron flow through the load resistor will also be spontaneous, driven by the net free energy change associated with the **cell reaction**. This is the essence of the electrochemical or **galvanic cell**, which is thus seen to be a device which allows the direct conversion of chemical into electrical energy.

The quantity of a chemical species reduced or oxidised in an electrochemical cell is related by **Faraday's Laws** to the total electric charge transferred across the metal–solution interfaces. A current of i amps flowing in the circuit for a time of δt seconds is equivalent to the transfer of $i.\delta t$ coulombs of charge across any interface in the cell. Now if the oxidation of each M^{x+}(aq) ion involves the transfer of one electron from the solution phase to the electrode (Fig. 2.1), then the passage of $i.\delta t$ coulombs must correspond to the oxidation of

$$i\, \delta t \Big/ (Le_0)$$

moles of M^{x+}(aq), where e_0 is the charge on an electron (1.601×10^{-19} C), and L is the Avogadro constant, or number of atoms in a mole (6.022×10^{23}). The term (Le_0) is given the name **Faraday's constant**, F, and has a

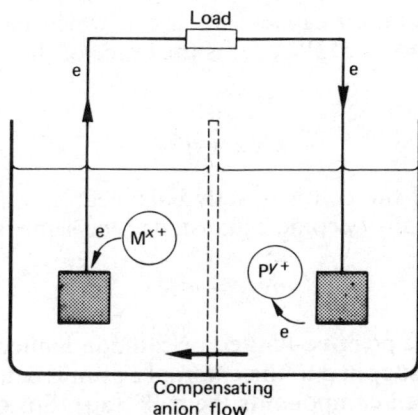

2.1 Basic electrochemical cell. Interconnection of the two electrodes through the external load resistor allows the spontaneous cell reaction to proceed.

value of $96\,490\,\text{C mol}^{-1}$. For the more general electrode process

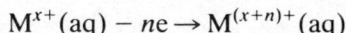

$$M^{x+}(aq) - ne \rightarrow M^{(x+n)+}(aq)$$

each $M^{x+}(aq)$ ion gives up n electrons to the external circuit in order to form the oxidised product. Here the number of moles of reactant $M^{x+}(aq)$ consumed by the passage of a current, i, flowing for a period δt is given by

$$N_M = \frac{i\,\delta t}{nF} \qquad (2.2)$$

Since the current passing through a cell is often a function of time, a more general form of equation 2.2 is

$$N_M = \frac{1}{nF}\int_o^t i\,dt \qquad (2.2')$$

where the integration is carried out over the interval during which current passes. Conversely, if N_M represents the total number of moles of $M^{x+}(aq)$ present, then the maximum **capacity** of the cell, that is the total amount of charge that could be supplied to the external circuit is given by

$$Q_T = \int_o^t i\,dt = nF.N_M \qquad (2.3)$$

Finally, recalling that current can only flow in a cell if there are two electrodes, the transfer of $i.\delta t$ coulombs in the appropriate direction at the second electrode (Fig. 2.1) brings about the reduction of $P^{y+}(aq)$ ions. If this second electrode process is

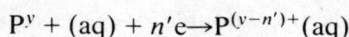

$$P^y + (aq) + n'e \rightarrow P^{(y-n')+}(aq)$$

then N_P, the number of moles of $P^{y+}(aq)$ reduced is given by

$$N_P = \frac{i\,\delta t}{n'F} \qquad (2.4)$$

Note that the maximum capacity of a cell which contains N_M moles of M^{x+}(aq) and N_P moles of P^{y+}(aq) is **the lesser** of the two quantities nFN_M and $n'FN_P$.

If

$$nN_M = n'N_P$$

the composition of the cell is exactly balanced, and the reactants at each electrode theoretically become exhausted at the same time.

In contrast, if

$$nN_M > n'N_P$$

the cell is said to be positive-limited or cathode-limited: i.e. the maximum charge that can be supplied to the external circuit is determined by only one of the two active cell components, here P^{y+}(aq). For operational reasons, most practical cells have such an imbalance in their make-up; this may be to prevent hazardous reactions such as the liberation of gases, or catastrophic failure caused by the puncture of a consumable container, at the end of the cell discharge.

2.2 The electrical double layer and the formation of electric potentials at interfaces

Whenever two phases which have significant electrical conductivities come into contact, there is generally some redistribution of charge – by charge transfer across the interface as described above, by adsorption of charged particles at the interface, by orientation of dipoles near the interface, etc. The region containing this distribution of charge is known as the **electrical double layer**, since the total charge excess on the one side of the interface must, for reasons of overall electroneutrality, be exactly equal and opposite to that on the other. When a metal is immersed in a solution of an electrolyte, the charge on the metal side of the interface corresponds to a surplus or deficit of electrons in the surface atomic layer. The structure of the solution side of the double layer is much more complicated: it can be considered to consist of two regions (Fig. 2.2(a)-(c)):

 (i) an **inner layer** which may contain partially desolvated anions which have strong chemical bonding interactions with the metal together with oriented solvent molecular dipoles, and

(ii) a 'time averaged' **diffuse layer** where ions are attracted to or repelled from the interface by electrostatic forces, but which are also affected by thermal collisions. The effective depth of the diffuse layer is a function of electrolyte concentration, temperature, dielectric constant of the solvent, etc.

   The formation of an electrical double layer at a metal–solution interface brings about a particular arrangement of ions in the region near the electrode surface and an associated variation in electric potential with distance from the interface. The double layer structure may affect significantly the rates of electrochemical reactions.

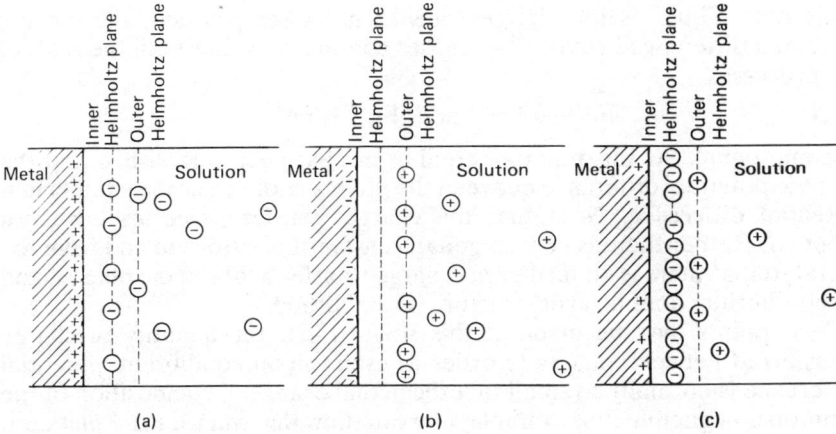

2.2 Structure of the electric double layer under different conditions of electrode polarisation:

(a) Metal positively charged, anions present at the inner Helmholtz plane (chemically interacting with metal) and in the diffuse double layer beyond the outer Helmholtz plane; (b) metal negatively charged, inner Helmholtz plane empty, cations in diffuse layer; (c) metal positively charged, strong adsorption of anions in inner Helmholtz plane, balancing cations in the diffuse layer.

In the case of electrochemical cells capable of passing significant direct currents, the principal mechanism for the formation of such potential differences at phase boundaries within the cell is **charge transfer**: e.g. electron transfer between two metals or semiconductors, ion transfer between a metal and a solution of its ions, etc. Consider a chemically inert metal, like platinum, immersed in a solution containing the oxidised and reduced forms of a chemical species, e.g. Fe^{3+} and Fe^{2+} ions. At the instant that the metal and solution are brought into contact, the two phases are uncharged and there is no electric potential difference between them. However, whenever they are brought into contact charge transfer processes begin: some electrons from the conduction band of the metal are accepted by Fe^{3+} ions in solution while some of the Fe^{2+} ions donate electrons to the metal. If the rates of these two processes are unequal, the metal and solution phases become progressively charged: for example, if the Fe^{3+} ions are more readily reduced than are the Fe^{2+} ions oxidised, the metal soon gains a net positive charge (through losing more electrons than it gains) and the solution an equal negative charge (made up of the excess balancing anions left over, when Fe^{3+} ions are converted to Fe^{2+} ions). As the electric charge on the two phases builds up, so an electrical potential difference develops between them. However the existence of such an electric potential difference affects the charge transfer rates: if the electrode (i.e. the metallic phase) acquires a relative positive charge, then the rate of reduction of Fe^{3+} ions will be depressed. At the same time, the rate of oxidation of Fe^{2+} ions will be increased. Thus, as the charge transfer processes proceed, the electric potential difference between the phases continues to alter until the rates of the processes

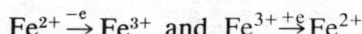

$$Fe^{2+} \xrightarrow{-e} Fe^{3+} \text{ and } Fe^{3+} \xrightarrow{+e} Fe^{2+}$$

become equal. At this point an equilibrium has been established and the electric potential difference between the phases is known as the **equilibrium potential difference**. Note that the charge transfer processes have not stopped. Rather there is now an equal transfer of electrons to and from the metal so that there is no further *net* charge transfer between the phases, and an equilibrium double layer structure is established.

Two points may be made at this stage. First, the quantity of charge transferred between phases in order to establish an equilibrium potential difference is normally so small that the actual change in composition of the solution is negligible. For example, one can show that when a 1 cm^2 platinum electrode is immersed in a Fe^{2+}/Fe^{3+} solution, a net reduction of between 10^{-9} and 10^{-10} moles of Fe^{3+} takes place. Second, and as will be stressed later, the kinetics of the charge transfer process are very important, since if rates are slow, it may not be possible for a true equilibrium to be established.

An essentially similar description can be given for the build up of a potential difference between a metal such as zinc and a solution containing Zn^{2+} ions, where the metal acts as the reduced form. Indeed a similar charge redistribution or double layer formation occurs whenever two different electric conductors are placed in contact. For instance when two different metals are connected together there will be a net flow of electrons from one to the other, resulting in a separation of charge between the two metallic phases and consequently what is known as a **contact potential difference** is set up.

The historically important Daniell cell consists of a copper vessel containing a saturated solution of copper sulphate which forms one of the two electrode systems, and a central amalgamated zinc rod immersed in a zinc sulphate/sulphuric acid solution, which forms the other (Fig. 2.3). Gross mixing of the two solutions is prevented by a membrane or porous pot. It is assumed here that both electrodes are provided with copper terminals.

In this cell there are three significant interfacial potential differences (neglecting any small liquid junction potential difference):

Cu	Zn	Solution	Cu
(1)	(2)	(3)	

The Galvani potential, ϕ of a phase defines the amount of electrical energy, $e\phi$ required to transport a charge e from an infinitely distant point in a vacuum to a hypothetical point in the interior of the phase where the charge would experience no "chemical" forces exerted on it. Thus the Daniell cell

Copper

Zinc

$ZnSO_4/H_2SO_4$ (aq)

Porous pot

$CuSO_4$ (aq)

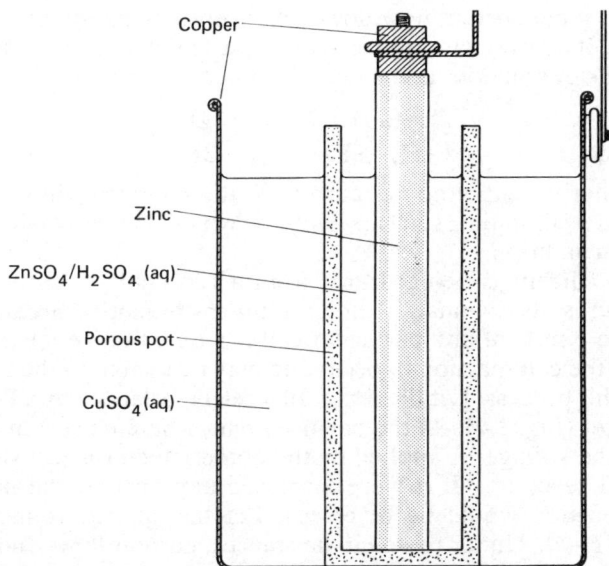

2.3   The Daniell cell.

voltage can be written as*

$$E_{cell} = {}^{Cu}\Delta^{Zn}\phi + {}^{Zn}\Delta^{Solution}\phi + {}^{Solution}\Delta^{Cu}\phi \qquad (2.5)$$

Unfortunately it is impossible to measure either the Galvani potential of a single phase or the Galvani potential difference between two phases of different composition. The only Galvani potential difference that can be measured in this cell is that existing between the two copper terminals. Thus it is not particularly instructive to develop further the concept of individual electric potential differences at phase boundaries within a galvanic cell. Instead, the relationship between the measurable **cell voltage** and the properties of the chemical processes associated with the running of the complete cell will now be considered.

2.3 Thermodynamics of galvanic cells

Energetics

Discharge of the Daniel cell causes zinc metal to ionise at one electrode and copper ions to deposit at the other. The net result of charge flow round the circuit is therefore equivalent to the reaction:

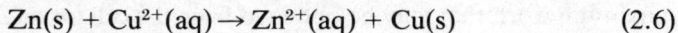

$$Zn(s) + Cu^{2+}(aq) \rightarrow Zn^{2+}(aq) + Cu(s) \qquad (2.6)$$

This is known as the **cell reaction**.

A cell is said to act reversibly when the net cell reaction is reversed when the current through the cell is made to flow in the opposite direction. When no current is being drawn, such a cell is in a true equilibrium state. Note that

---

* Note that ${}^{Cu}\Delta^{Zn}\phi$ is by no means negligible: from work function measurements it can be shown to have a value of about 0.25V.

the absence of net current flow does not necessarily signify that a cell is in equilibrium. If an iron wire is placed in a solution of low pH the most likely electron transfer reactions at the metal/solution interface are

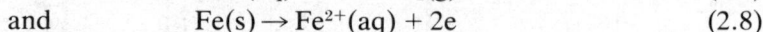

$$2H^+(aq) + 2e \rightarrow H_2(g) \tag{2.7}$$

and $$Fe(s) \rightarrow Fe^{2+}(aq) + 2e \tag{2.8}$$

At a particular potential no net current at the interface will be observed. However a cell containing such an electrode would obviously fail to fulfil the reversibility condition.

When no current is being drawn from a reversible cell, the potential difference across its terminals is known as the **electromotive force** or **e.m.f.** of the cell. The e.m.f. of any particular cell is a quantitative measure of the tendency of the cell reaction to occur and may be related to the free energy change for this process. Let the e.m.f. of a cell be balanced by a Poggendorf potentiometer (Fig. 2.4). At the balance point, where no current is flowing in the cell, the voltage, E applied by the potentiometer is just sufficient to stop the cell reaction. (If it were increased any further, the normal cell discharge reaction would be reversed). Let the applied voltage now be reduced to $(E-\delta)$. Under these circumstances, current flows and the work

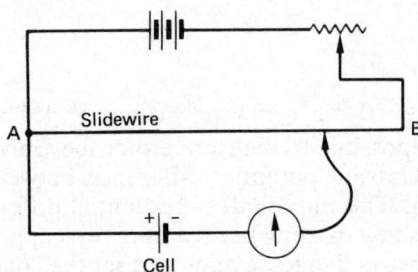

2.4 The Poggendorf potentiometer: a circuit which permits the e.m.f. of a cell to be backed-off so that the spontaneous cell reaction can be allowed to proceed, to be exactly halted (galvanometer at zero), or to be reversed.

done, in joules, by the cell as one mole of reactants are converted to products would be equal to the product of the charge driven through the applied voltage (say nF coulombs), and the value of this potential difference $(E - \delta$ volts):

i.e. $$w = nF(E - \delta) \tag{2.9}$$

n is the number of moles of electrons transferred in one mole of reaction. Obviously the value of δ can be made as small as desired, so that the maximum work that may be obtained from the cell for one mole of reaction is

$$w_{max} = nFE \tag{2.10}$$

By definition, $w_{max} = -\Delta G$, where ΔG is the free energy change associated with one mole of reaction and hence

$$\Delta G = -nFE \tag{2.11}$$

In practice, passage of any finite amount of electricity results in a certain degree of irreversibility as some of the electrical energy is dissipated as Joule heat in the internal resistance of the cell, or at the electrodes where application of a potential difference is required to drive the current at the desired rate. Galvanic cells can only supply electric work equal to the free energy change of the cell reaction when the current flowing tends to zero. The cell voltage under load is always smaller than the e.m.f. so that only part of the thermodynamically available work can be utilised. The various causes of internal resistance or 'polarisation' of cells are considered later.

The enthalpy change associated with a cell reaction is a state function and hence is independent of whether the reaction is being carried out reversibly or not. If an exothermic reaction is carried out completely irreversibly, the total enthalpy change ΔH will appear as heat given out to the surroundings:

$$\text{Heat given out} = q_{out} = -\Delta H \tag{2.12}$$

On the other hand if some work is done by the cell (e.g. by turning an electric motor in order to lift a weight) then

$$w_{out} + q_{out} = -\Delta H \tag{2.13}$$

The Second Law of Thermodynamics determines how much work can be extracted from a process: for maximum work output there is an associated minimum heat output.

$$(w_{out})_{max} + (q_{out})_{min} = -\Delta H \tag{2.14}$$

The term $(q_{out})_{min}$ may be identified with $-T\Delta S$ where ΔS is the entropy change of the cell reaction. If ΔS is positive the cell will cool down as it operates, or take in heat from the surroundings. If the same cell operates irreversibly, it will cool down less, or take in less heat from the surroundings.

Schematic representation of a galvanic cell

A galvanic cell may be represented by writing the composition of the individual phases of which it is composed in the order in which they are connected. A convenient notation is to indicate each phase boundary by a vertical line. The zone of contact of two solution phases is indicated by a double vertical line. (This region is usually associated with a small non-equilibrium potential difference caused by charge separation resulting from the unequal rates of diffusion of ions across it.) The accepted convention is for a cell diagram to be drawn so that the cell reaction under consideration, whether spontaneous or not, takes place when positive charge flows through the cell from left to right. The cell voltage is then equal in sign and magnitude to the voltage of the right hand terminal, taking that of the left as zero. E_{cell}, the cell e.m.f., is positive when the cell reaction proceeds spontaneously on connecting the two terminals together. Thus if m_1 and m_2 refer to the molal concentrations of $Zn^{2+}(aq)$ and $Cu^{2+}(aq)$ respectively it is equally correct to write for the Daniell cell either

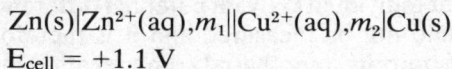

$$Zn(s)|Zn^{2+}(aq),m_1||Cu^{2+}(aq),m_2|Cu(s)$$
$$E_{cell} = +1.1 \text{ V}$$

which describes the spontaneous discharge process

$$Zn(s) + Cu^{2+}(aq),m_2 \rightarrow Zn^{2+}(aq),m_1 + Cu(s)$$

or

$$Cu(s)|Cu^{2+}(aq),m_2\|Zn^{2+}(aq),m_1|Zn(s)$$
$$E_{cell} = -1.1 \text{ V}$$

which describes the non-spontaneous charging reaction

$$Zn^{2+}(aq),m_1 + Cu(s) \rightarrow Zn(s) + Cu^{2+}(aq),m_2$$

In practice the first of these alternatives is generally used. The implication of this is that galvanic cells are written with the positive electrode on the right, electron flow in the external circuit is from left to right and in the solution phase, cation flow is from left to right and anion flow from right to left (Fig. 2.5).

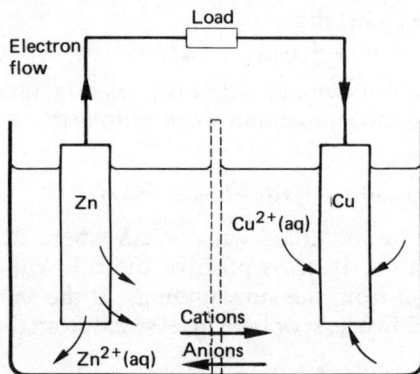

2.5　Ion and electron flow in a schematic Daniell cell.

It is convenient at this point to introduce the terms **anode** and **cathode**. A cathode is defined as the electrode to which electrons flow from the external circuit – i.e. as the electrode at which reduction takes place. Conversely, electrons flow from the anode to the external circuit: oxidation takes place at the anode. The point of difficulty rests in the fact that in a galvanic cell under spontaneous discharge the cathode is the positive electrode, whereas in an electrolytic cell or galvanic cell under charge, it is the negative electrode (Fig. 2.6).

Half-reactions, half-cells and electrode potentials

The e.m.f. of a reversible cell can be regarded either as a function of the free energy change associated with the overall cell reaction or as a sum of the Galvani potential differences between phases within the cell. It was noted above, however, that individual Galvani potential differences between non-identical phases could not be measured and it is therefore impossible to resolve a cell e.m.f. into its interphasial components. On the other hand,

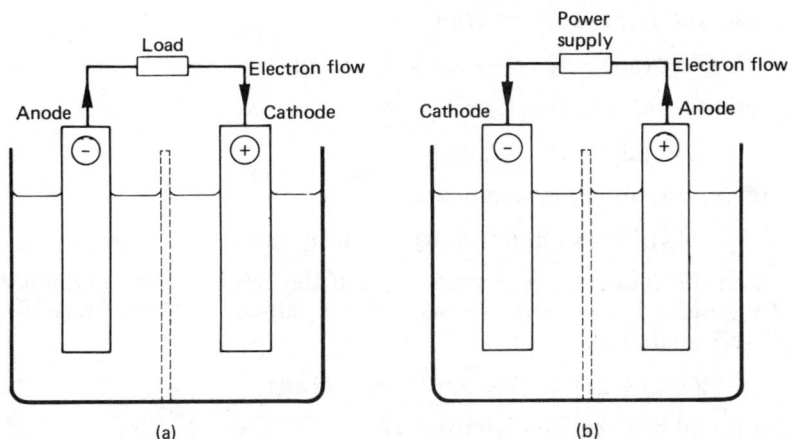

2.6   Anode and cathode in a cell during (a) spontaneous discharge and (b) during charge.

every cell **reaction** consists of an oxidation and a reduction process and thus can be considered as the sum of two notional 'half-reactions' occurring in notional 'half-cells'. For example, the Daniell cell can be visualised as consisting of the half-cells

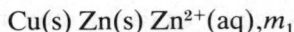

$$Cu(s)\ Zn(s)\ Zn^{2+}(aq),m_1$$

and $$Cu(s)\ Cu^{2+}(aq),m_2$$

If cells are constructed by making different combinations of half-cells, then the cell e.m.f. values obey an additive law. It is therefore convenient to combine half-cells with a single reference half-cell and thus obtain a series of related e.m.f. values compared to the reference half-cell taken as zero.

The universally accepted primary reference half-cell is the standard hydrogen electrode. The electrode consists of a noble metal (platinised platinum) dipping into a solution of hydrogen ions at unit activity and saturated with hydrogen gas at one atmosphere of pressure. (In practice such a **standard** electrode cannot be realised, but the scale it defines can.)

The **electrode potential** is defined as the potential difference between the terminals of a cell constructed of the half-cell in question and a standard hydrogen electrode (or its equivalent) and assuming that the terminal of the latter is at zero volts. Note therefore that the electrode potential is an observable physical quantity and is unaffected by the conventions used for writing cells. The statement '... the electrode potential of zinc is -0.76 volts ...' implies only that a voltmeter placed across the terminals of a cell consisting of standard hydrogen electrode and the zinc electrode would show this value of potential difference with the zinc terminal negative with respect to that of the hydrogen electrode. An electrode potential is **never** a 'metal/solution potential difference', not even on some arbitrary scale.

An alternative way of presenting data on cells with a standard hydrogen electrode is in terms of the **reduction potential** or **oxidation potential** of the

cell reaction. For the cell reaction

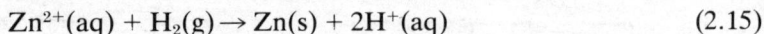

$$Zn^{2+}(aq) + H_2(g) \rightarrow Zn(s) + 2H^+(aq) \qquad (2.15)$$

the notional half-reaction is written as

$$Zn^{2+}(aq) + 2e \rightarrow Zn(s) \qquad (2.16)$$

and the e.m.f. of the corresponding cell

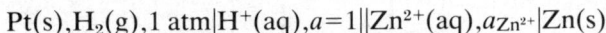

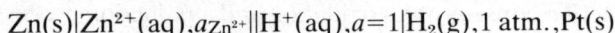

$$Pt(s),H_2(g),1 \text{ atm}|H^+(aq),a=1\|Zn^{2+}(aq),a_{Zn^{2+}}|Zn(s)$$

is called the reduction potential, E_{Red}. If the cell reaction (and hence the half-reaction) is reversed, the sign of the half-cell potential must also be reversed, so that for

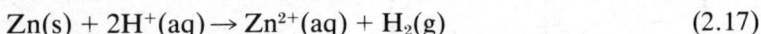

$$Zn(s) + 2H^+(aq) \rightarrow Zn^{2+}(aq) + H_2(g) \qquad (2.17)$$

the notional half-reaction is written as

$$Zn(s) \rightarrow Zn^{2+}(aq) + 2e \qquad (2.18)$$

and the e.m.f. of the corresponding cell

$$Zn(s)|Zn^{2+}(aq),a_{Zn^{2+}}\|H^+(aq),a=1|H_2(g),1 \text{ atm.},Pt(s)$$

is defined as the oxidation potential, E_{Ox}. Thus

$$(E_{Ox})_{Zn} = -(E_{Red})_{Zn} \qquad (2.19)$$

It is perhaps best to consider oxidation and reduction potentials as properties of a half-reaction rather than of a half-cell. (Some confusion might be avoided if tables of reduction potentials were written in terms of the complete cell reaction rather than as notional half-reactions). It can be seen that the electrode potential has the same value as the reduction potential of the half-reaction. For the Daniell cell, the following relationships exist:

$$Zn(s)|Zn^{2+}(aq)\|Cu^{2+}(aq)|Cu(s)$$

E.m.f. of cell $= E_{cell}$
$\qquad$ = Electrode potential of copper – electrode potential of zinc
$\qquad = (E_{Red})_{Cu} - (E_{Red})_{Zn}$
$\qquad = (E_{Red})_{Cu} + (E_{Ox})_{Zn}$
$\qquad = (E_{Ox})_{Zn} - (E_{Ox})_{Cu} \qquad (2.20)$

Rules for the manipulation of reduction (or oxidation) potentials

(i) A half-reaction can be multiplied through by any number without affecting the reduction (or oxidation) potential.

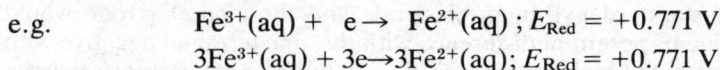

e.g. $\qquad Fe^{3+}(aq) + e \rightarrow Fe^{2+}(aq)$; $E_{Red} = +0.771$ V
$\qquad\qquad 3Fe^{3+}(aq) + 3e \rightarrow 3Fe^{2+}(aq)$; $E_{Red} = +0.771$ V

(ii) Subtraction of reduction (or of oxidation) potentials of two half-reactions having the same number of electrons gives the e.m.f. of the corresponding cell reaction.

e.g.

$$Cl_2(g) + 2e \rightarrow 2Cl^-(aq) \qquad ; E_{Red} = +1.360 \text{ V}$$

$$Zn^{2+}(aq) + 2e \rightarrow Zn(s) \qquad ; E_{Red} = -0.763 \text{ V}$$

∴ for $\quad Cl_2(g) + Zn(s) \rightarrow 2Cl^-(aq) + Zn^{2+}(aq) ; E_{cell} = +2.123 \text{ V}$

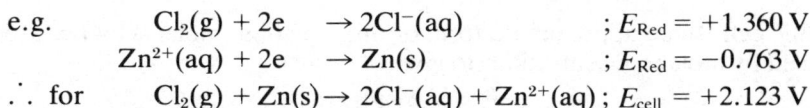

(iii) Subtraction of reduction (or of oxidation) potentials of two half-reactions not having the same number of electrons does **not** give the reduction (or oxidation) potential of the resulting new half-reaction.

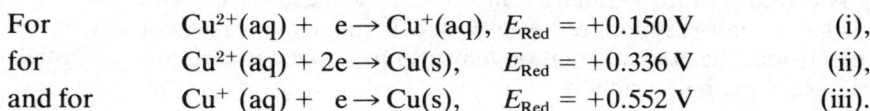

For $\qquad Cu^{2+}(aq) + e \rightarrow Cu^+(aq), E_{Red} = +0.150 \text{ V}$ (i),

for $\qquad Cu^{2+}(aq) + 2e \rightarrow Cu(s), \qquad E_{Red} = +0.336 \text{ V}$ (ii),

and for $\quad Cu^+ (aq) + e \rightarrow Cu(s), \qquad E_{Red} = +0.552 \text{ V}$ (iii).

Subtracting half-reaction (i) from half-reaction (ii) gives half-reaction (iii); subtracting the reduction potentials gives $+0.187$ V which is **not** the reduction potential of (iii).

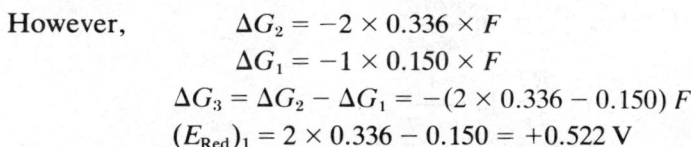

However, $\qquad \Delta G_2 = -2 \times 0.336 \times F$

$$\Delta G_1 = -1 \times 0.150 \times F$$

$$\Delta G_3 = \Delta G_2 - \Delta G_1 = -(2 \times 0.336 - 0.150) F$$

$$(E_{Red})_1 = 2 \times 0.336 - 0.150 = +0.522 \text{ V}$$

Dependence of e.m.f. on concentration

The dependence of e.m.f. on the activities or concentrations of the products and reactants of a cell reaction follows directly from a consideration of the relationship between e.m.f. and free energy change. For the cell reaction

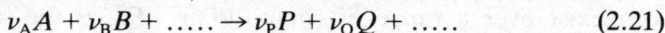

$$v_A A + v_B B + ... \rightarrow v_P P + v_Q Q + ... \qquad (2.21)$$

if v_i is the stoichiometric number and μ_i is the chemical potential of a species involved, then

$$\Delta G = \sum_i v_i \mu_i = \sum_i v_i (\mu_i^{\ominus} + RT \ln a_i) \qquad (2.22)$$

Here $\mu_i^{\ominus}$ is the chemical potential of i in its standard state and a_i is its molal activity. This equation can be recast in the form known as the van't Hoff reaction isotherm:

$$\Delta G = \Delta G^{\ominus} + RT \ln \left\{ \frac{a_P{}^{v_P} a_Q{}^{v_Q} ...}{a_A{}^{v_A} a_B{}^{v_B} ...} \right\} \qquad (2.23)$$

Hence the e.m.f. of a cell is given by

$$E = E^{\ominus} - \frac{RT}{nF} \ln \left\{ \frac{a_P{}^{v_P} a_Q{}^{v_Q} ...}{a_A{}^{v_A} a_B{}^{v_B} ...} \right\} \qquad (2.24)$$

$E^{\ominus}$ is the **standard e.m.f. of the cell** and is the equilibrium voltage when all the cell components are in their standard states: solution species have unit molal activities, gases have pressures of 1 atmosphere and solid phases are in their most stable form. Equation 2.24 is known as the Nernst equation for a galvanic cell. In an approximate form of this equation, molal activities are replaced by molar concentrations to give

$$E = E^{\ominus} - \frac{RT}{nF} \ln \left\{ \frac{[P]^{\nu_P} [Q]^{\nu_Q} \cdots}{[A]^{\nu_A} [B]^{\nu_B} \cdots} \right\} \qquad (2.24')$$

(The values of the standard e.m.f. in 2.24 and 2.24' are slightly different.) Since a half-reaction represents a cell involving a standard hydrogen electrode, the Nernst equation may also be applied to electrode potentials. Thus for the half-reaction

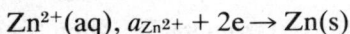

$$Zn^{2+}(aq), a_{Zn^{2+}} + 2e \rightarrow Zn(s)$$

the electrode potential is given by

$$E = E_{Zn}^{\ominus} - \frac{RT}{2F} \ln \frac{a_{Zn(s)}}{a_{Zn^{2+}}}$$

$$= E_{Zn}^{\ominus} - \frac{RT}{2F} \ln \frac{1}{a_{Zn^{2+}}}$$

$$= E_{Zn}^{\ominus} + \frac{RT}{2F} \ln a_{Zn^{2+}}$$

$$\approx E_{Zn}^{\ominus} + 0.0295 \log_{10} [Zn^{2+}] \text{ at } 25° \text{ C.} \qquad (2.25)$$

Note that $a_{Zn(s)} = 1$ since zinc is a pure substance. A more complex relationship is required if one component is a non-stoichiometric solid which can exist over a range of compositions. $E_{Zn}^{\ominus}$ is the **standard electrode potential of zinc.**

The dependence of cell e.m.f. on the concentrations of reactants and products of the cell reaction is of fundamental importance in the understanding and design of practical battery systems. As a cell undergoes discharge, reactants are steadily converted to products until one of the reactants species is completely exhausted. Consider the cell reaction $A + B \rightarrow C + D$, where A–D are all solution species. For this system,

$$E_{cell} = E_{cell}^{\ominus} - \frac{RT}{nF} \ln \left\{ \frac{a_C \, a_D}{a_A \, a_B} \right\}$$

At low discharge levels the concentration of reactants will be large and that of the products small, so that E_{cell} will be relatively high. As the discharge continues, the thermodynamic-controlled open circuit voltage (i.e. the cell e.m.f.) will rapidly fall. If a battery is to be designed to have only a small decrease in open circuit voltage from the initial open circuit value across the entire discharge, it is best to select reactants and products whose activities remain effectively constant. This is the situation when these components are pure materials or the major constituent in a phase whose composition does not vary significantly. In Fig. 2.7 and 2.8 the open circuit voltage is plotted as a function of percentage discharge for the cells

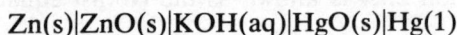

$$Zn(s)|ZnO(s)|KOH(aq)|HgO(s)|Hg(1)$$

for which the cell reaction is

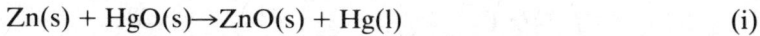

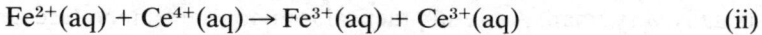

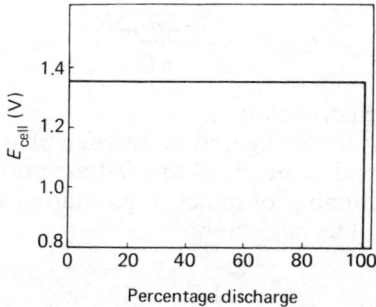

$$Zn(s) + HgO(s) \rightarrow ZnO(s) + Hg(l) \qquad \text{(i)}$$

and

$$Pt(s)|Fe^{2+}(aq),Fe^{3+}(aq)\|Ce^{3+}(aq),Ce^{4+}(aq)|Pt(s)$$
$$\quad 0.1M \quad\;\; 0.1M \qquad\;\; 0.1M \quad\;\; 0.1M$$

for which the cell reaction is

$$Fe^{2+}(aq) + Ce^{4+}(aq) \rightarrow Fe^{3+}(aq) + Ce^{3+}(aq) \qquad \text{(ii)}$$

2.7 Open circuit voltage as a function of percentage discharge for the cell Zn(s)|ZnO(s)|KOH(aq)|HgO(s)|Hg(l).

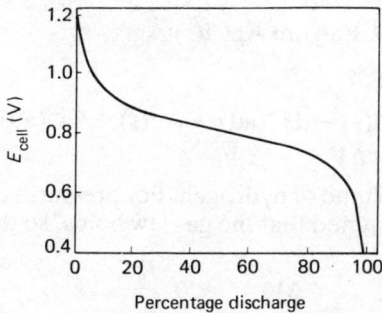

2.8 Open circuit voltage as a function of percentage discharge for the cell Pt(s)|Fe²⁺(aq), 0.1 mol dm⁻³, Fe³⁺(aq), 0.1 mol dm⁻³ ‖ Ce³⁺(aq), 0.1 mol dm⁻³, Ce⁴⁺(aq), 0.1 mol dm⁻³|Pt(s).

Dependence of e.m.f. on temperature and pressure

From the identity

$$dG = -SdT + Vdp$$

it is seen that

$$\left(\frac{\partial \Delta G}{\partial T}\right)_p = -\Delta S \quad \text{or} \quad \left(\frac{\partial E}{\partial T}\right)_p = \frac{\Delta S}{nF} \qquad (2.26)$$

and

$$\left(\frac{\partial \Delta G}{\partial p}\right)_T = \Delta V \quad \text{or} \quad \left(\frac{\partial E}{\partial p}\right)_T = \frac{-\Delta V}{nF} \tag{2.27}$$

Cell e.m.f. values may therefore have positive or negative temperature coefficients, depending on the sign of the entropy change associated with the cell reaction. It should be noted that unless gases are involved in the latter, the entropy change and hence the temperature coefficient of the e.m.f. is usually very small. A useful practical form for the temperature dependence is

$$E_T = E_{298} + \frac{(T-298)\Delta S}{nF} \tag{2.28}$$

where T is the Celsius temperature.*

Again for cell reactions involving only condensed phases, the e.m.f. can be regarded as virtually independent of applied pressure. However when there is a change in the number of moles of gas during a cell reaction the volume change, ΔV cannot be neglected:

$$E_p = E_{p^{\ominus}} - \frac{1}{nF} \int_{p^{\ominus}}^{p} \Delta V \mathrm{d}p \tag{2.29}$$

where $p^{\ominus}$ represents the standard pressure. For a cell such as

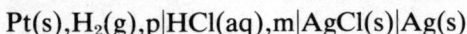

$$\mathrm{Pt(s),H_2(g),p|HCl(aq),m|AgCl(s)|Ag(s)}$$

for which the cell reaction is

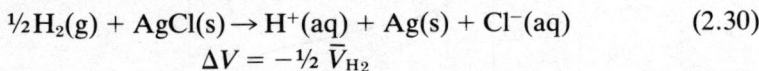

$$\tfrac{1}{2}\mathrm{H_2(g)} + \mathrm{AgCl(s)} \rightarrow \mathrm{H^+(aq)} + \mathrm{Ag(s)} + \mathrm{Cl^-(aq)} \tag{2.30}$$
$$\Delta V = -\tfrac{1}{2} \bar{V}_{\mathrm{H_2}}$$

where $\bar{V}_{\mathrm{H_2}}$ is the molar volume of hydrogen. For pressures up to some tens of atmospheres it can be assumed that the gas law holds, so that

$$\Delta V = -\tfrac{1}{2} \frac{RT}{p}$$

Hence
$$E_p = E_{p^{\ominus}} + \frac{RT}{2nF} \int_{p^{\ominus}}^{p} \frac{\mathrm{d}p}{p}$$

$$= E_{p^{\ominus}} + \frac{RT}{2nF} \ln p \tag{2.31}$$

At higher pressures more accurate equations of state are required and ΔV generally becomes a more complex function of the pressure.

---

* It is well known that the working voltage of a practical cell under load usually rises with an increase in temperature. However this is almost entirely due to a reduction in the internal resistance of the cell caused by an increase in the conductivity of the electrolytic phase and in the diffusion rates of the electroactive species.

Electrode types

Electrodes or half-cells are often classified into three main types. The classification is mainly historical since there are no fundamental differences between them.

(i) Electrodes of the first kind

Electrodes of this form have a minimum of two phases, namely a metal (either in pure form, or in an amalgam or solid solution) in contact with an electrolyte phase (solution, melt, solid electrolyte, glass, etc.) containing cations of the metal.

For the half-reaction

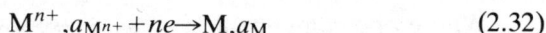

$$M^{n+}, a_{M^{n+}} + ne \rightarrow M, a_M \tag{2.32}$$

the electrode potential is given by

$$E = E_{M^{n+}/M}^{\ominus} + \frac{RT}{nF} \ln \frac{a_{M^{n+}}}{a_M} \tag{2.33}$$

If the phase containing M is pure metal then $a_M = 1$.

(ii) Electrodes of the second kind

In these systems there are a minimum of three phases: a metal, an adherent layer of sparingly soluble salt of the metal, and an electrolyte phase containing the same anion as the latter. Equilibrium exists both between the metal and cations of the sparingly soluble salt and between anions in the salt and those in solution. For the half-reaction

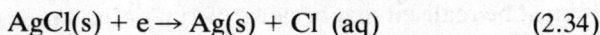

$$AgCl(s) + e \rightarrow Ag(s) + Cl^-(aq) \tag{2.34}$$

the electrode potential is given by

$$
\begin{aligned}
E &= E_{Ag^+/Ag}^{\ominus} + \frac{RT}{F} \ln a_{Ag^+} \\
&= \left(E_{Ag^+/Ag}^{\ominus} + \frac{RT}{F} \ln K_s \right) - \frac{RT}{F} \ln a_{Cl^-} \\
&= E_{Ag/AgCl/Cl^-}^{\ominus} - \frac{RT}{F} \ln a_{Cl^-} \tag{2.35}
\end{aligned}
$$

where K_s is the thermodynamic solubility product of AgCl.

The most common electrode of this type is the saturated calomel electrode (SCE) which consists of mercury in contact with a layer of insoluble Hg_2Cl_2 immersed in a saturated aqueous solution of KCl. The SCE is used as a *secondary standard* reference electrode. At 25°C it has an electrode potential of +0.2415 V.

(iii) Redox electrodes

In redox electrodes an inert metal conductor acts as a source or sink for electrons. The components of the half reaction are the two oxidation states of a constituent of the electrolytic phase. Examples of this type of system include the ferric/ferrous/electrode where the active components are cations, the ferricyanide/ferrocyanide electrode where they are anionic complexes, the hydrogen electrode, the chlorine electrode, etc. In the gaseous electrodes equilibrium exists between electrons in the metal, ions in solution and dissolved gas molecules. For the half-reaction

$$Ox(aq) + ne \rightarrow Red(aq) \tag{2.36}$$

the electrode potential is given by

$$E = E^{\ominus}_{Ox/Red} + \frac{RT}{nF} \ln \frac{a_{Ox}}{a_{Red}} \tag{2.37}$$

Where one component is a gas, it is assumed that equilibrium exists between the gas phase and gas molecules dissolved in solution so that the activity of the latter is given by the partial pressure in the gas phase. Thus for

$$\tfrac{1}{2}Cl_2(g), p_{Cl_2} + e \rightarrow Cl^-(aq), a_{Cl^-} \tag{2.38}$$

$$E = E^{\ominus}_{Cl_2/Cl^-} + \frac{RT}{F} \ln \frac{(p_{Cl_2})^{1/2}}{a_{Cl^-}} \tag{2.39}$$

Note that homogeneous electron transfer in the solution phase is negligible in the case of half-reactions involving gases, but that it may be very rapid for ionic redox systems.

Indirect values of electrode potentials

It must be realised that because of kinetic limitations most half-cells that can be written cannot be the basis of a practical cell which will display the appropriate e.m.f. It has however proved convenient to include such half equations in tables of redox potentials if their e.m.f. could be evaluated in some other way. **In a large number of cases electrochemical data are not used at all.** Rather, partial molar heats and entropies of the species involved are determined by calorimetric methods and these are used to derive $\Delta G^{\ominus}$ for the cell reactions. $E^{\ominus}_{cell}$ values can then be calculated.

For example, consider how the standard potential for the process

$$2CO_2(g) + 2H^+(aq) + 2e \rightarrow H_2C_2O_4(aq)$$

might be found. This half-cell reaction implies the cell reaction

$$2CO_2(g) + 2H^+(aq) \xrightarrow{2e} H_2C_2O_4(aq)$$
$$H_2(g) \xrightarrow{-2e} 2H^+(aq)$$

i.e.
$$2CO_2(g) + H_2(g) \rightarrow H_2C_2O_4(aq)$$

If the free energies of formation of oxalic acid and carbon dioxide are known,

then
$$\Delta G^{\ominus} = (\Delta G^{\ominus}_f)_{H_2C_2O_4} - 2(\Delta G^{\ominus}_f)_{CO_2} - 0$$
$$= (-695.0) - 2 \ (-395.3)$$
$$= + 95.6 \, \text{kJ mol}^{-1}$$

and hence
$$E^{\ominus}_{cell} = \Delta G^{\ominus}/nF = -0.495 \, \text{V}$$

Liquid junction potentials

When a cell contains a boundary between two electrolytic solutions of different composition or concentration, a liquid junction potential is developed due to the diffusion of the various components at characteristic rates in the boundary zone. Cells with liquid junction potentials cannot be regarded as reversible or equilibrium systems: rather, a constant liquid junction potential may be interpreted as indicating a steady state where an effectively time-independent charge separation has been developed. Diffusion, however, tends to be a slow process, so that the structure of the zone may hardly change. This is even more true when a **separator** or porous insulating matrix is used to immobilise the solution boundary in the boundary, or where gelling agents are added to the electrolytes. Values of the liquid junction potential are difficult to calculate except for special cases – e.g. for the boundary between solutions having different concentrations of the same components. Liquid junction potentials are highest when there is a large concentration difference between the phases or where cation and anion mobilities are notably different. Under these circumstances liquid junction potentials of up to 50 mV may occur. However in practical battery systems, any liquid junction potential is likely to be considerably lower and is almost always neglected.

2.4 Current flow in an electrochemical cell

Relationship between current and reaction rate

Within any cell undergoing charge or discharge, one can consider at least three forms of charge transmission:
(i)   electron flow in the electronic conductors (e.g. electrode materials, terminal connectors, load resistor, etc.),
(ii) ion flow in the electrolyte (which may be an aqueous solution, solid electrolyte, molten salt, etc.), and
(iii) charge transfer reactions at the electrode/electrolyte interfaces (e.g. $Zn(s) \xrightarrow{-2e} Zn^{2+}(aq)$).

Since in the steady state, it is necessary to maintain a condition of electroneutrality in any macroscopic part of the system, the total charge flux through all cross sections of the circuit must be the same. In particular,

**Rate of electron flow in external circuit
= rate of charge transfer at each electrode/electrolyte interface.**

For the general electrode process

$$Ox + ne \rightarrow Red$$

nF coulombs are required to reduce one mole of Ox, so that the rate of reduction of Ox at the electrode is given by

$$\left\{ \frac{\text{Rate of passage of electrons into electrode from external circuit}}{nF} \right\}$$

i.e. Rate of reduction of $Ox = i/nF$ mol s^{-1} (2.40)

At the other electrode, $Red' - n'e \rightarrow Ox'$

and again

$$\text{Rate of oxidation of } Red' = i/n'F \text{ mol s}^{-1}$$ (2.40')

It should be noted that the rate of charge transfer at the electrodes (i.e. the rates of the electrochemical processes) may be given directly by the reading on an ammeter inserted in the external circuit. If the current is a function of time it is still possible to apply the above idea of 'flux continuity' over a succession of small time intervals. Now it may happen that one of the various rate processes involved in charge transport in different components of a cell, is unable to maintain as high a rate as the others. Under these circumstances it becomes the **current limiting** process for the cell.

The question now arises as to what factors are responsible for determining the rates at which the various cell processes occur. Thermodynamic arguments permit the feasibility of overall cell reactions to be predicted, but give no information on rates. To understand the latter it is necessary to consider the effects on various parts of the cell of forcing the cell voltage to assume a value different from that of the equilibrium e.m.f. It has been shown above that in the Daniell cell at equilibrium, charge transfer across the zinc/solution interface can be described in terms of processes

$$Zn^{2+}(aq) \xrightarrow{2e} Zn(s)$$

and $$Zn(s) \xrightarrow{-2e} Zn^{2+}(aq)$$

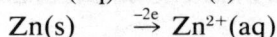

occurring at equal rates. The first of these may be considered as constituting a cathodic flow of charge across the interface: i.e. as a cathodic current, i_c. Similarly for the oxidation of metallic zinc to zinc ions, there is a charge flow which may be described as an anodic current, i_a. At equilibrium i_c is equal to i_a and this equilibrium current is known as the **exchange current**, i_o. The net current i, given as $(i_c - i_a)$ by convention, is of course zero at equilibrium. In the same way, at the copper electrode at equilibrium,

$$i_c' = i_a' = i_o'$$

(Note that it would be most improbable for i_o and i_o' to be equal.) Now let a constant voltage source be placed across the cell so that the zinc is forced to assume a less negative potential and the copper a less positive potential than their equilibrium values. The altered electric field across the zinc/solution interface makes it easier for zinc atoms to be oxidised but hinders the reduction of zinc ions. Hence i_a increases from its equilibrium value while i_c decreases: therefore

$$|i_{Zn}| = i_c - i_a \neq 0.$$

At the copper electrode, reduction of Cu^{2+} is favoured and oxidation of Cu atoms is restricted, so that net cathodic flow occurs. Finally, to prevent a build up of Zn^{2+} ions near the zinc/electrolyte interface and of SO_4^{2-} counter ions near the copper, a flux of ions must take place in the electrolytic phase to balance the charge transfer processes at the interfaces. To maintain the flux continuity condition, the applied voltage difference becomes distributed in such a way that*

$$|i_{Zn}| = |i_{Cu}| = |i_{\text{ion flow}}| \qquad (2.41)$$

For spontaneous discharge, the overall cell voltage must be reduced from its equilibrium value, as would happen if a load resistor were connected to the terminals. If a potential difference greater than the e.m.f. were applied (i.e. one making the cathode more positive and the anode more negative) the net result would be a current flow in the reverse direction, causing a net charging of the cell.

Polarisation losses

Knowledge of the amount by which the voltage of a cell, delivering a particular level of current, deviates from its equilibrium value is of central importance in assessing the performance of a practical battery system. This polarisation voltage, E_{pol}, can be associated with two principal causes:

(i) 'ohmic' or 'iR' drop in the bulk of the electrolyte phases, separators and sometimes in the electrode phases and connectors, and

(ii) 'electrode losses' which include the 'activation overvoltage', connected with the charge transfer step and/or nucleation and crystallisation processes at each electrode/electrolyte interface, and the 'concentration overvoltage' related to the depletion or accumulation of electroactive material near the electrode surfaces.

In practical batteries, especially those employing porous phases, it is not always possible to separate ohmic and electrode losses clearly.

Note that for cell discharge,

$$E_{\text{cell}} = E_{\text{thermodynamic}} - E_{\text{pol}} \qquad (2.42)$$

while for a cell undergoing charge,

$$E_{\text{cell}} = E_{\text{thermodynamic}} + E'_{\text{pol}} \qquad (2.43)$$

---

* The voltage becomes distributed just as in the case of a potential, E, applied to three resistors in series:

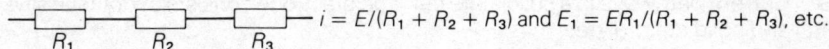

$i = E/(R_1 + R_2 + R_3)$ and $E_1 = ER_1/(R_1 + R_2 + R_3)$, etc.

Ohmic potential drop

The source of ohmic potential drop is the internal resistance of the bulk phases within the cell. If the current distribution is uniform, then for a phase with conductance σ (Sm^{-1}), the resistance is $R = x/A\sigma$ where x is the thickness of the phase and A its cross-sectional area. Thus for the passage of a current i through a cell with j sequential phases,

$$E_{ohmic} = i \sum_j (x_j/A_j\sigma_j) \tag{2.44}$$

In practice, current distribution rarely approaches a uniform distribution. Instead it is a complex function of cell geometry, characteristics of the electrode surfaces, etc. It is sometimes possible to determine experimentally an effective or average value of R_{ohmic} using transient perturbation techniques. However in many situations it is difficult to distinguish ohmic and electrode polarisation, for instance where the electroactive material is also responsible for carrying all the current in the electrolyte phase, in the situation where porous electrodes are involved, or where electrode and electrolyte phases are finely mixed, as in some solid-state cells. In Fig. 2.9 two different conduction pathways are shown in a cell where the electrode material is granular silver ground together with a silver ion conducting solid electrolyte powder, and compressed to minimise electrode polarisation by forming a large interfacial contact area. A very similar situation occurs with a porous electronically conducting electrode substrate and a liquid solution electrolyte.

It is common in many practical battery designs to 'immobilise' a liquid electrolyte phase within a porous solid insulator. The electrolyte conductivity and ohmic loss in such a system is determined by the number of pores, their size, shape and tortuosity. The tortuosity coefficient, β, is

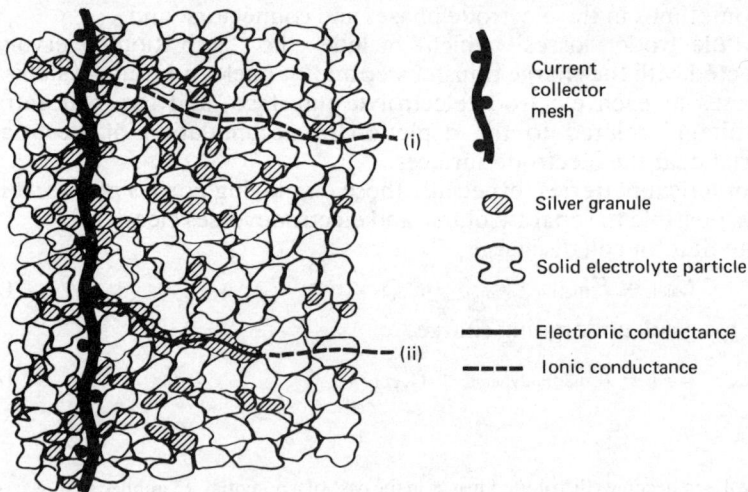

Current collector mesh

Silver granule

Solid electrolyte particle

————— Electronic conductance

- - - - - Ionic conductance

2.9   Current pathways in a solid state cell: the electrode consists of granular silver metal and solid electrolyte.

defined as the ratio of the mean distance covered by an ion traversing a porous matrix, to the direct distance of one side of the matrix to the other. The relative reduction in the conductivity of an electrolyte solution caused by confining it in a porous solid is called the **conductivity attenuation, Θ**. For a matrix of uniform cylindrical pores it is given by

$$\Theta = \pi N_p r_p^2 / \beta \tag{2.45}$$

where N_p is the number of pores (with radius r_p) per unit area of the solid matrix, and

$$\kappa_M = \Theta \kappa \tag{2.46}$$

where κ is the conductance of the free electrolyte solution and κ_M is its conductance when immobilised.

In most modern practical batteries, a major part of polarisation loss at moderately high current densities is due to ohmic potential drop. Considerable attention is therefore given during the design of a battery

(a)   to obtaining maximum ionic conductance in the electrolyte phases (a particular problem in certain non-aqueous and solid electrolyte cells),

(b)   to reducing the resistivities of electrode materials (especially when they are in extended form, such as sintered powders, etc.) and supplying adequate current collectors, inter-cell connectors, etc.,

(c)   to balancing the advantages from an electrode polarisation viewpoint of high interfacial areas against the drawbacks of high resistance paths in the bulk phases, and

(d)   to optimising cell geometry so as to minimise the effective distance between electrodes, while maintaining uniform current distribution.

This last consideration, namely uniform current distribution, is frequently of critical importance since variable current densities and the associated non-uniform potential distribution can result in localised depletion of reactants, shape changes in electrodes, side reactions, low electrode material utilisation, etc. A simple example is shown in Fig. 2.10 where the potential distribution in a cylindrical cell with unequal electrode areas is shown. Very high current densities may occur near the circumference of the smaller electrode. As has been pointed out above, current flow in a cell is determined by (i) ohmic loss in the electrolyte and (ii) electrode polarisation losses at the electrode/electrolyte interfaces. If the latter is high, then the effect of uneven ohmic potential gradients becomes unimportant. However under normal circumstances, ohmic loss is substantial and great care must therefore be taken during the design stage to try to establish uniform current

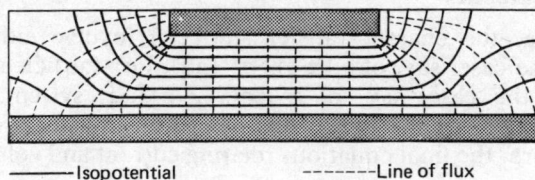

———— Isopotential   - - - - - - - Line of flux

2.10   Potential distribution in a cylindrical cell with electrodes of unequal area.

densities at the electrode surfaces. In practice the current density is affected by both potential and concentration gradients, so that theoretical treatments have been confined to rather simple models.

Electrode kinetics and electrode polarisation losses

Polarisation losses associated with an electrode process are termed **overvoltage**. The overvoltage η is defined as

$$\eta = E - E_{eq} \tag{2.47}$$

where E_{eq} is the equilibrium (zero current) potential of an electrode and E is its potential when a certain current is being passed. Since an overall electrode process can be subdivided into a number of consecutive steps, involving transport of the electroactive species from the bulk of the electrolyte phase to the electrode surface, interfacial charge transfer, etc., a number of different effects may contribute to the total overvoltage. Of fundamental importance is the interfacial charge transfer step whose rate is directly controlled by the potential difference across the double layer and hence by the electrode potential. 'Charge transfer' is considered here to include the whole process of chemical bond formation and scission, solvation changes, etc. which must accompany the successful transfer of an electron or ion across an electrode/solution interface. In addition to the basic interfacial charge transfer, many electrode processes involve a further step in the vicinity of the phase boundary – e.g. electrocrystallisation, in which atoms produced in the charge transfer step are incorporated into the lattice, or some slow chemical reaction, whose rate is independent of the electrode potential. Finally all electrode processes must involve mass transport whereby materials consumed or formed in the charge transfer step are moved to or from the electrode surfaces. Mechanisms of mass transport may include diffusion, convection and electromigration.

In some circumstances, one of the possible contributing factors to the overvoltage is dominant and it is possible to say that the electrode process is, for instance, 'mass transport' or 'charge transfer' limited. In other cases, more than one step is responsible for significantly restricting the current flow and overvoltage contributions from all such steps must be taken into account. Polarisation losses of various kinds can sometimes be distinguished by subjecting the electrode to current or voltage transients (e.g. step or sinusoidal functions) and analysing the response.

Charge transfer kinetics

As noted above, the charge transfer step can involve either electron transfer, as in the case, say, of a $Pt(s)|Fe^{3+}(aq)$, $Fe^{2+}(aq)$ electrode or ion transfer, as in the case, say, of a $Zn(s)|Zn^{2+}(aq)$ system. While the theoretical treatment of the two forms of charge transfer are essentially different in nature, the final equations relating current and voltage turn out to be very similar.

In the case of redox electrodes, the ease with which electrons can tunnel through a potential barrier of the type present at an electrode interface makes the use of classical activated complex theory (with the electrons as one reactant) inappropriate. In Fig. 2.11 (a) an electron energy diagram of a redox electrode at equilibrium is shown. For an electron transfer between the phases to be successful, it is necessary for the acceptor or donor in solution to have an energy level exactly equal to a complementary level in the metal. In the equilibrium situation it is seen that there is an equal chance of transfer of an electron from a filled metal level to an unoccupied solution level of the same energy or from an occupied solution level to an empty metal level. By making the electrode potential more negative (Fig. 2.11 (b)), the Fermi level of the metal is raised, more filled electronic states in the metal overlap with unfilled states in the solution and hence there is a relatively greater tendency for electron transfer from the metal to the solution. The opposite situation occurs when the electrode potential is made more positive (Fig. 2.11 (c)). The current flowing at a redox electrode is not limited by the rate at which electrons can tunnel to or from the metal: rather it is restricted by the rates at which the solution species can undergo thermal fluctuations to produce transient species (e.g. ions with abnormal hydration sheaths, or molecules with stretched bonds) which have suitable electronic states. For the electrode process

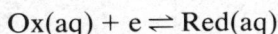

$$Ox(aq) + e \rightleftharpoons Red(aq)$$

the forward and reverse reactions can be considered as normal heterogeneous rate processes, but with rate constants which are potential dependent. Thus

$$i_c = k_c FA [Ox], \tag{2.48}$$
$$i_a = k_a FA [Red] \tag{2.48'}$$

where k_c and k_a are heterogeneous rate constants, A is the area of the electrode/solution interface and $[Ox]$ and $[Red]$ are the surface concentrations of the electroactive species – here assumed to be the same as the bulk values. The rate constants have the form

$$k_c = k_0 \exp \left\{ -\alpha F(E - E^\ominus)/RT \right\} \tag{2.49}$$

and

$$k_a = k_0 \exp \left\{ (1-\alpha)F(E - E^\ominus)/RT \right\} \tag{2.50}$$

where E is the potential of the electrode, $E^\ominus$ is its **standard** equilibrium potential and k_o is the single heterogeneous rate constant characteristic of the redox system at the temperature of interest. α is a constant between zero and one, known as the **transfer coefficient** or **symmetry factor**: its value is generally close to 0.5.

For the electrode to be in equilibrium,

$$i_0 = i_c \qquad (= i_a)$$
$$= k_0 FA [Ox] \exp \left\{ -\alpha F(E - E^\ominus)/RT \right\}$$
$$= k_0 FA [Ox] \exp \ln \left\{ \frac{[Ox]^{-\alpha}}{[Red]^{-\alpha}} \right\}$$

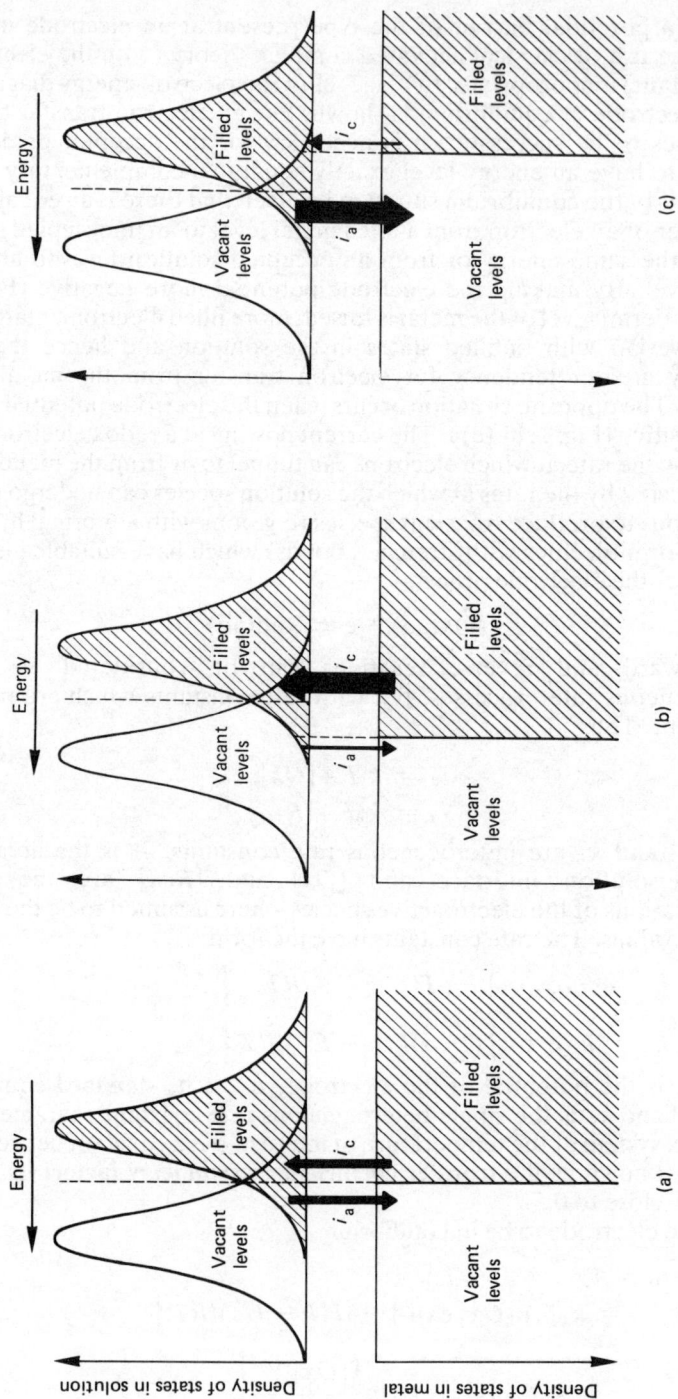

2.11 Occupation of energy levels in a metal and a solution redox couple.

$$= k_oFA \ [Ox]^{(1-\alpha)} \ [Red]^\alpha \tag{2.51}$$

This general expression for the exchange current reduces to

$$i_o = k_oFA \, [Ox] = k_oFA \, [Red] \tag{2.52}$$

for the situation where $[Ox] = [Red]$, i.e. at the standard potential of the couple, and to

$$i_o = k_oFA \tag{2.53}$$

for the special case corresponding to the situation where

$$[Ox] = [Red] = 1.$$

Whenever there is a net flow of current, this is taken by convention as

$$i = i_c - i_a.$$

For $i_c \neq i_a$, it is no longer strictly correct to equate the surface and bulk concentrations of the electroactive species. However if the charge transfer rate is current limiting, then as a first approximation it can be assumed that the surface concentration does not deviate significantly from the bulk value. Hence

$$i = k_oFA \left[\delta \, [Ox] \exp \left\{ -\alpha F(E - E^\ominus)/RT \right\} - [Red] \exp \left\{ (1-\alpha)F(E - E^\ominus)/RT \right\} \right]$$

Also $\eta = E - E_{eq}$, so that

$$(E - E^\ominus) = \eta + \frac{RT}{F} \ln \frac{[Ox]}{[Red]}$$

Hence on substituting for $(E - E^\ominus)$ and recalling the expression derived above for i_o,

$$i = i_o \left\{ \exp \frac{-\alpha F \eta}{RT} - \exp \frac{(1-\alpha)F\eta}{RT} \right\} \tag{2.54}$$

This equation is known as the Butler-Volmer relationship. Fig. 2.12 shows that the net current flowing at an electrode can be considered as the resultant or sum of cathodic and anodic contributions. Note that in equation 2.54 and in subsequent equations dealing with electrode kinetics, the current is a direct function of the electrode area, A. It is often useful to normalise such equations with respect to area by replacing the current i by the **current density**, i/A.

The exact shape of the curve is defined by the two parameters k_o and α. In Fig. 2.13, α is held constant at 0.5 while k_o is varied. It can be seen that if the heterogeneous rate constant is high, large currents can be generated by very small overvoltages. In Fig. 2.14 the variation of α for a constant k_o is seen to alter the symmetry of the current/voltage behaviour. An interesting feature of this figure is the coincidence of all the curves near the origin: use can be made of this for determining i_o. The **polarisation resistance**, R_p, of the electrode is defined as the inverse of the current/potential gradient at the

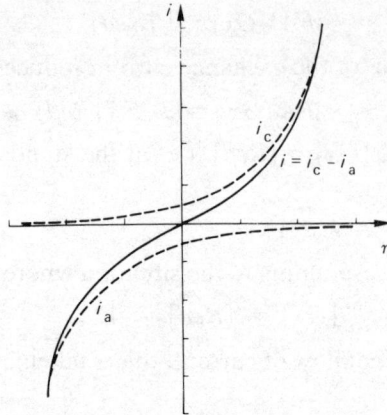

2.12 Net current flow, i, at an electrode as the resultant of cathodic and anodic contributions, i_c and i_a.

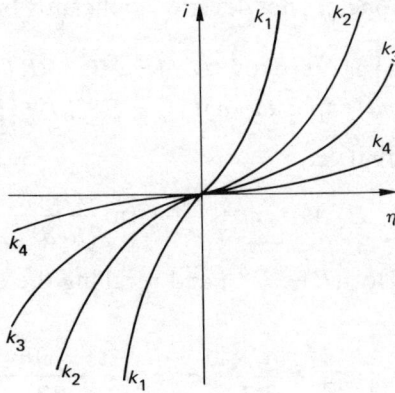

2.13 Influence of k_0 on the shape of the current-potential curve, for constant α.
$$k_1 > k_2 > k_3 > k_4.$$

origin: i.e.

$$R_p = 1/(di/d\eta)_{\eta=0} \tag{2.55}$$

Now $\qquad \dfrac{di}{d\eta} = i_0 \left\{ \dfrac{-\alpha F}{RT} \exp \dfrac{-\alpha F\eta}{RT} - \dfrac{(1-\alpha)F}{RT} \exp \dfrac{(1-\alpha)F\eta}{RT} \right\}$

$\therefore \qquad \left(\dfrac{di}{d\eta}\right)_{\eta=0} = i_0 \left\{ \dfrac{-\alpha F}{RT} - \dfrac{(1-\alpha)F}{RT} \right\} = \dfrac{-i_0 F}{RT}$

Hence $\qquad\qquad\qquad R_p = -\ RT/i_0 F \tag{2.56}$

i_o can therefore be found by taking the tangent to the current/potential curve at the origin. In practice, when η is small, the general expression for the current can be simplified by using the relationship:

$$\exp(-x) \approx 1 - x \qquad \text{for } x \text{ small.}$$

Under these circumstances,

$$i = - i_o F\eta/RT \tag{2.57}$$

Another construction for obtaining i_o is suggested later.

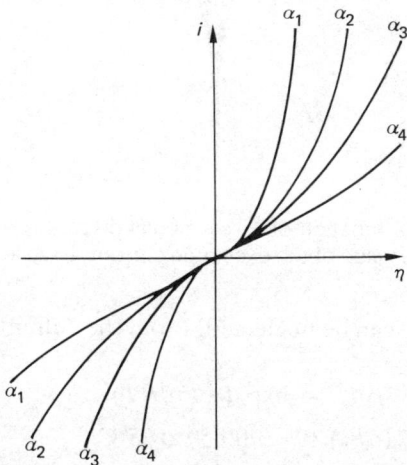

2.14 Influence of α on the shape of the current-potential curve, for constant k_o.

$$\alpha_1 > \alpha_2 > \alpha_3 > \alpha_4.$$

There are a number of methods available for determining α. For instance rearranging equation 2.51 gives

$$i_o/[Ox] = k_o FA([Ox]/[Red])^{-\alpha}, \text{ so that}$$
$$\log i_o/[Ox] = \log k_o FA - \alpha \log [Ox]/[Red]$$

Hence a plot of $\log i_o/[Ox]$ versus $\log[Ox]/[Red]$ gives α from the slope. The transfer coefficient may also be evaluated using a form of the Tafel equation as outlined below.

If α and i_o are known, then k_o is given as

$$k_o = (i_o/FA) [Ox]^{(1-\alpha)} [Red]^{\alpha} \tag{2.58}$$

in the general case; or

$$k_o = (i_o/FA) [Ox] = (i_o/FA)[Red] \tag{2.59}$$

at the standard potential of the couple; or

$$k_o = i_o/FA \tag{2.60}$$

where $\qquad [Ox] = [Red] = 1$

For large absolute values of η, only that component of the electrode process favoured by the direction of the overpotential need be considered:

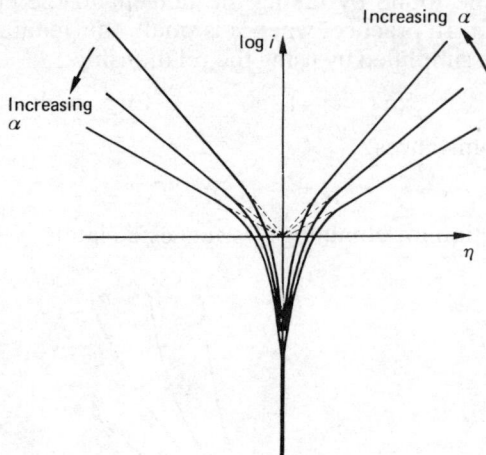

2.15 Plots of log *i* as a function of η: extrapolation of the linear portions of the curves to $\eta = 0$ gives the value of the exchange current i_0, while α may be evaluated from their gradients.

i.e. the 'back reaction' can be neglected. From the definitions of i_c and i_a we have

$$i_c/i_a = \exp\,(-F\eta/RT) \qquad (2.61)$$

or $\qquad \log_{10}(i_c/i_a) = -\eta/0.059$ at 25°C.

Thus for an overvoltage of -118 mV, $i_c/i_a = 100/1$. For overvoltages of 200 mV and above, only the cathodic or anodic contribution to the total current need be considered. For a large cathodic overvoltage.

$$i = i_c - i_a$$
$$\approx i_c$$
$$= i_0 \exp\,(-\alpha F\eta/RT)$$
$$\therefore \qquad \ln i = \ln i_0 - \alpha F\eta/RT \qquad (2.62)$$

Plots of log i/i_0 as a function of η are shown in Fig. 2.15. By extending the linear portions of these curves to $\eta = 0$, a value of the exchange current is given, while the α coefficients may be evaluated from their gradients. Equation 2.62 may be recast in the form:

$$\eta = (2.303RT/\alpha F)\log_{10}i_0 - (2.303RT/\alpha F)\log_{10}i \qquad (2.62')$$

or as $\qquad \eta = A + B \log_{10} i,$

which is the usual form of the well-known **Tafel equation**. A similar expression exists for anodic currents.

The electric field which actually affects the charge transfer kinetics is that between the electrode and the plane of closest approach of the solvated electroactive species ('outer Helmholtz plane'), as shown in Fig. 2.2. While the potential drop across this region generally corresponds to the major component of the polarisation voltage, a further potential fall occurs in the 'diffuse double layer' which extends from the outer Helmholtz plane into the bulk of the solution. In addition, when ions are specifically adsorbed at the electrode surface, Fig. 2.2(c), the potential distribution in the inner part of the double layer is no longer a simple function of the polarisation voltage. Under these circumstances serious deviations from Tafel-like behaviour are common.

Many redox reactions at electrodes involve transfer of more than one electron. It is agreed that such processes usually involve several consecutive one-electron steps rather than a simultaneous multi-electron transfer. The kinetics of the overall reaction (and hence the current flowing) are complicated by such factors as the lifetimes of the transient intermediate species.

The theoretical model generally used for predicting the overvoltage-current function for metal/metal ion systems is based on the quasi-thermodynamic arguments of transition state theory. The anodic charge transfer process is considered to involve the rupture of the bond between an 'adatom' – i.e. a metal atom in a favourable surface site – and the metal, followed by, or coincident with, the formation of electrostatic bonds between the newly formed ion and solvent or other complexing molecules. The cathodic charge transfer follows this mechanism in reverse.

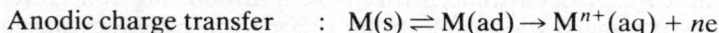

Anodic charge transfer　:　$M(s) \rightleftharpoons M(ad) \rightarrow M^{n+}(aq) + ne$

Cathodic charge transfer　:　$M^{n+}(aq) + ne \rightarrow M(ad) \rightleftharpoons M(s)$

It is assumed that for pure charge transfer current limitation, the equilibrium between $M(s)$ and $M(ad)$ is fast and potential independent. Under these conditions the relevant equations are

$$k_a = k_0 \exp\left\{ (1-\alpha)nF(E-E^\ominus)/RT \right\} \tag{2.63}$$

$$i_a = k_0 nFA\,[M(ad)]\exp\left\{ (1-\alpha)nF(E-E^\ominus)/RT \right\} \tag{2.64}$$

$$k_c = k_0 \exp\left\{ -\alpha nF(E-E^\ominus)/RT \right\} \tag{2.65}$$

$$i_c = k_0 nFA\,[M^{n+}(aq)]\exp\left\{ -\alpha nF(E-E^\ominus)/RT \right\} \tag{2.66}$$

$$i_0 = k_0 nFA[M^{n+}(aq)]^{(1-\alpha)}[M(ad)]^\alpha$$

and

$$i = i_0 \left\{ \exp\frac{-\alpha nF\eta}{RT} - \exp\frac{(1-\alpha)nF\eta}{RT} \right\} \tag{2.68}$$

The close similarity between equations 2.68 and 2.62 can be seen. Equation 2.68 leads in the same way to Tafel behaviour at high overvoltages.

Concentration overvoltage

The term **concentration overvoltage** is used to describe restrictions in the current caused by concentration changes of the electroactive species at the electrode surface. In practical batteries such an effect is only important at relatively high current densities. If attention is focused first on the electrolytic phase, it is seen that the supply of electroactive molecules or ions to the electrode surface can be maintained either by some process of mass transport from the bulk of the phase or by a (potential independent) chemical reaction occurring in the electrolytic phase near the interface. The principal mechanisms of mass transport in batteries are diffusion in a concentration gradient and migration in an electric field. If ions are present in the electrolyte which are not involved in the electrode process, then transport by electromigration may be small. On the other hand, when the electroactive materials are responsible for carrying all the current in the electrolyte phase, it is not possible to distinguish mass transport polarisation and ohmic potential drop. For an electrode process $Ox + ne \rightarrow Red$, concentration overvoltage is given by

$$
\begin{aligned}
\eta &= E - E_{eq} \\
&= \left\{ E^{\ominus} + \frac{RT}{nF} \ln \frac{[Ox]_o}{[Red]_o} \right\} - \left\{ E^{\ominus} + \frac{RT}{nF} \ln \frac{[Ox]}{[Red]} \right\} \\
&= \frac{RT}{nF} \ln \left\{ \frac{[Ox]_o}{[Ox]} \cdot \frac{[Red]}{[Red]_o} \right\}
\end{aligned}
\tag{2.69}
$$

where $[Ox]_o$ and $[Red]_o$ represent the concentrations of electroactive species adjacent to the electrode surface as the result of passing a particular level of current.

In the case of mass transport by pure diffusion, the concentrations of electroactive species at an electrode surface can often be calculated for simple systems by solving **Fick's equations** with appropriate boundary conditions. A well known example is for the overvoltage at a planar electrode under an imposed constant current and conditions of semi-infinite linear diffusion. The relationships between concentration, distance from the electrode surface, x, and time, t, are determined by solution of Fick's Second Law so that expressions can be written for $[Ox]_o$ and $[Red]_o$ as functions of time. Thus for

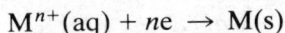

$$
M^{n+}(aq) + ne \rightarrow M(s)
$$

and assuming (i) that the activity of the deposited metal remains at unity, and (ii) that the surface layer is not completely depleted of $M^{n+}(aq)$ ions, then

$$
[M^{n+}]_o = [M^{n+}] - \frac{2it^{1/2}}{nFAD^{1/2}\pi^{1/2}}
\tag{2.70}
$$

where D is the diffusion coefficient of $M^{n+}(aq)$ and i is the constant impressed current. Further,

$$
E = E^{\ominus} + \frac{RT}{nF} \ln [M^{n+}]_o \text{ and } E_{eq} = E^{\ominus} + \frac{RT}{nF} \ln [M^{n+}]
$$

so that

$$
\eta = \frac{RT}{nF} \ln \frac{nFAD^{1/2}\pi^{1/2}[M^{n+}] - 2it^{1/2}}{nFAD^{1/2}\pi^{1/2}[M^{n+}]}
\tag{2.71}
$$

However in most practical situations it is not possible to obtain analytical expressions of this type. A useful model, originally intended for systems with a stirred electrolytic phase, assumes that a thin 'stationary layer' extends some way from the electrode surface into the bulk electrolytic phase and that beyond this layer, no significant concentration gradients exist

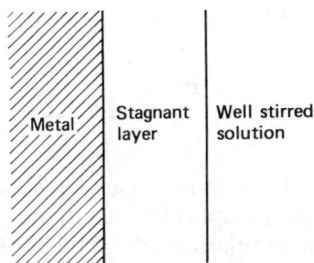

2.16   Stationary layer model.

(Fig. 2.16). This model predicts that after polarisation for an initial period, a time-invariant overvoltage will be established. This model has been applied with some success to unstirred systems where it is thought that natural convection caused by density variations (arising from changes in the composition of the electrolyte due to the electrode reaction) permits a similar layer system to be established. Such a model is not applicable where the electrolyte is immobilised, nor is it useful for electrodes with rough or porous surfaces.

Again for the process $M^{n+}(aq) + ne \rightarrow M(s)$ and assuming first that diffusion is the only mechanism of transport across the stationary layer, in the steady state, the flux of electroactive species across the layer must be balanced by the flux of charge at the electrode surface. For a linear concentration gradient across a stationary layer of thickness δ, application of Fick's First Law gives

$$i = nFAD \left\{ ([M^{n+}] - [M^{n+}]_0)/\delta \right\} \tag{2.72}$$

The form of this equation shows that the concentration gradient can have any value from zero up to a maximum of $[M^{n+}]/\delta$ where the surface concentration of $M^{n+}(aq)$ has fallen to zero. This maximum concentration gradient is associated with a maximum or **limiting** current given by

$$i_{lim} = nFAD\,[M^{n+}]/\delta \tag{2.73}$$

Note that it is straightforward to correct equations such as 2.72 and 2.73 for electromigration. Thus

$$i = nFAD \left\{ \frac{[M^{n+}] - [M^{n+}]_0}{\delta(1 - t_{Mn+})} \right\} \tag{2.74}$$

and

$$i_{lim} = nFAD \left\{ \frac{[M^{n+}]}{\delta(1 - t_{Mn+})} \right\} \tag{2.75}$$

where t_{Mn+} is the transport number of $M^{n+}(aq)$ in solution.

The diffusion overvoltage according to this model can be evaluated using the relationship

$$[M^{n+}]_o = (1 - i/i_{lim})[M^{n+}] \tag{2.76}$$

so that
$$\eta = \frac{RT}{nF} \ln \frac{i_{lim} - i}{i_{lim}} \tag{2.77}$$

It should be remembered that in the case of a metal/metal ion electrode, the current can only be mass-transport limited in the cathodic direction. For net anodic currents, an accumulation of metal ions occurs at the electrode surface. The anodic current can greatly exceed the limiting cathodic current (Fig. 2.17) and under these circumstances equations can be simplified to give

$$\eta = \frac{RT}{nF} \ln \frac{|-i|}{i_{lim}} \tag{2.78}$$

where i refers to the anodic current and i_{lim} to the limiting cathodic current. Hence

$$\eta = \left\{ \frac{RT}{nF} \ln i_{lim} \right\} - \frac{RT}{nF} \ln i \tag{2.79}$$

This equation has exactly the same form as the Tafel equation: a linear Tafel plot does not necessarily imply a charge-transfer limited process.

In the models for mass transport considered so far, it has been assumed that the electrode surface was smooth. In practice rough or porous structures are commonly used in battery systems to increase the 'real' surface area and so reduce the charge transfer overvoltage. When surface irregularities are of the same order of magnitude as the thickness of the diffusion layer δ, calculation of diffusion flux becomes very difficult. With porous electrodes the real surface area may be $10^3 - 10^5$ times greater than the projected or geometric surface. However the presence of a pore structure imposes fundamental changes both on charge-transfer kinetics because of the altered potential distribution and on mass-transport control because of the effect of the pores on concentration gradients. A number of simple models have been studied (e.g. a sequence of parallel cylindrical pores), and while there are no general solutions, a number of limiting cases have been described quantitatively.

In addition to mass transport from the bulk of the electrolyte phase, electroactive material may also be supplied at the electrode surface by homogeneous or heterogeneous chemical reaction. For example, hydrogen ions required in an electrode process, may be generated by the dissociation of a weak acid. As this is an uncommon mechanism so far as practical batteries are concerned (but not so for fuel cells) the theory of reaction overvoltage will not be further developed here. However it may be noted that Tafel-like behaviour and the formation of limiting currents are possible in reaction controlled electrode processes.

The discussion of concentration polarisation so far has centred on the depletion of electroactive material on the electrolyte side of the interface. If the metal deposition and dissolution processes involve metastable active

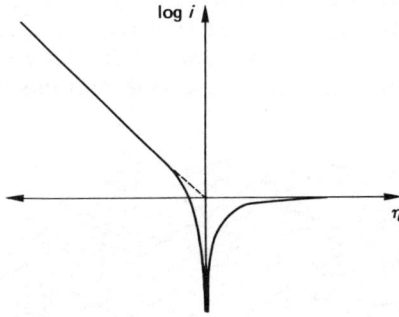

2.17 Current-voltage curve for a metal–metal ion electrode.

surface atoms, then the rate of formation or disappearance of these may be the critical factor in the overall electrode kinetics. Equation 2.69 can be rewritten for 'crystallisation overvoltage' as

$$\eta = \frac{RT}{nF} \ln \frac{[M(ad)]_{eq}}{[M(ad)]} \qquad (2.80)$$

where $[M(ad)]_{eq}$ and $[M(ad)]$ represent the surface concentrations (or, better, activities) of M adatoms on the metal surface under conditions of equilibrium and current flow respectively. Some features associated with electrocrystallisation and electrodissolution are shown in Fig. 2.18. Atoms such as A are the adatoms: they may be considered as the species formed or

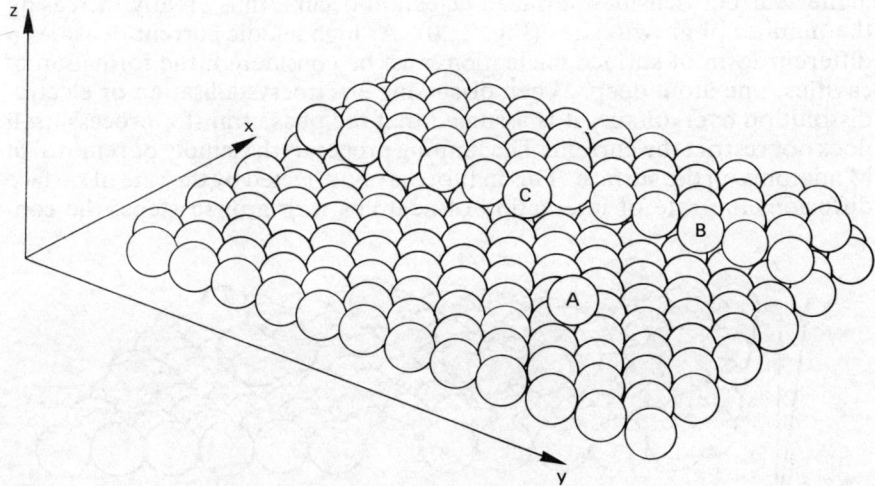

2.18 Electrocrystallisation on a metal surface. Upper layer of atoms grows by surface diffusion of **adatoms** (A) across plane XY until they encounter **step site** of lower energy, such as B. The adatom may now move along the step until it is located at the even lower energy **kink site** at C.

lost during charge transfer. (In other theories, adatoms are considered to have undergone only partial charge transfer.) Adatoms are free to move about on surfaces such as XY by an activated process of surface diffusion. Atoms at B and C are seen to exist at different positions of lower energy known as step sites.

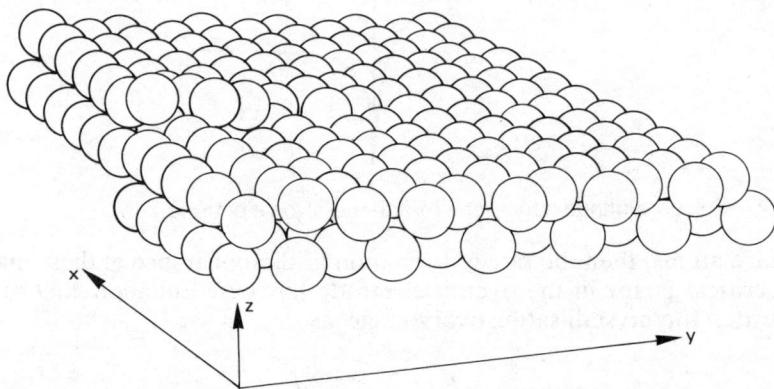

2.19 Electrocrystallisation on a metal surface. Growth can proceed continuously in the Z-direction as the step rotates around the screw dislocation.

At low current densities, electrocrystallisation is thought to proceed by incorporation of adatoms in step sites. The steps may then grow, and provided that each step is located at a screw dislocation, Fig. 2.19, the number of step sites remains constant during deposition or dissolution. At higher current densities surface nucleation occurs, thus greatly increasing the number of growth sites (Fig. 2.20). At high anodic current densities a different form of surface nucleation must be considered: the formation of cavities, one atom deep. When discussing electrocrystallisation or electro-dissolution overvoltage, it is assumed that the phase transfer process itself does not restrict the current. The limiting process is the supply or removal of M adatoms on the surface. This in turn may be affected by the rate of surface diffusion, the rate of interaction of adatoms with growth steps, the con-

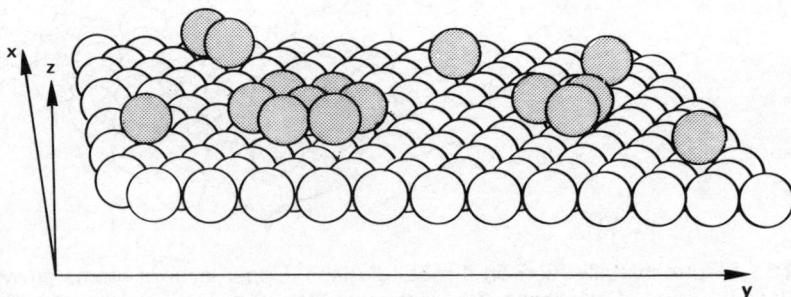

2.20 Electrocrystallisation on a metal surface. At high concentrations of adatoms two-dimensional surface nuclei may form which provide growth steps.

centration and distribution of nuclei, etc. The overvoltage may therefore follow a wide variety of current functions. If surface diffusion is rate-limiting, then in the simplest cases the overvoltage/current relationship becomes indistinguishable from that of charge transfer. This is also the case for sufficiently large rates of surface nucleus formation. Where both nucleation and screw dislocation growth proceed simultaneously, the current/voltage relationship is complex.

It is therefore seen that equations 2.42 and 2.43 for the on-load cell voltages during charge and discharge respectively can be expanded to give:

$$E_{cell} = E_{thermodynamic} - E_{pol} \tag{2.42}$$

$$= E_{thermodynamic} - \eta_{cathode} - \eta_{anode} - iR \tag{2.42'}$$

and $\quad E_{cell} = E_{thermodynamic} + E'_{pol} \tag{2.43}$

$$= E_{thermodynamic} + \eta'_{cathode} + \eta'_{anode} + iR \tag{2.43'}$$

The cathodic and anodic overvoltages depend on the dominant rate limiting processes for a particular current density and direction.

2.5 Battery characteristics and performance criteria

A cell may be characterised in terms of (i) its available capacity, (ii) its available energy, and (iii) the power it can deliver. It is not very useful to consider efficiency in the sense used to assess heat engines since batteries cannot be regarded as thermal converters. Rather, the efficiency of an electrochemical power source can be expressed in terms of capacity or of energy delivered.

Capacity

The **theoretical capacity** of a cell or half-cell may be calculated as

$$Q_T = x\,(nF) \tag{2.81}$$

where x is the theoretical number of moles of reaction associated with the complete discharge of the cell. The **practical capacity** or actual number of coulombs or ampere hours delivered, Q_P, is lower than Q_T if utilisation of electroactive material is not 100% due, for instance, to some chemical reaction occurring in the cell which consumes some of the reactants. The **rated capacity** is the practical capacity of a cell which has been discharged under prescribed conditions until the cell voltage has fallen to a pre-selected **cut-off voltage**.

The **coulombic efficiency** of a cell is defined as Q_P/Q_T. It is often more useful to determine the capacity of each half-cell separately, since for operational reasons, most practical batteries do not have an equal number of equivalents of anodic and cathodic reactants.

For purposes of comparison, it is convenient to calculate the **specific capacity**, defined as the capacity divided by the mass of the cell or half-cell, and usually given in units of Ah kg^{-1}. In some cases a volume-based specific capacity is preferred (e.g. Ah dm^{-3}).

Energy

The **theoretical available energy** for one mole of reaction (not for complete discharge) is given by

$$\mathscr{E}_T = -\Delta G = nFE_{cell} \qquad (2.82)$$

where E_{cell} is the e.m.f. The actual amount of energy delivered for one mole of reaction, or **practical available energy** is

$$\mathscr{E}_P = \int_0^{nF} E.dq = \int_0^t (Ei)dt \qquad (2.83)$$

and is dependent on the manner in which the cell is discharged. The units of energy are either joules (i.e. watt seconds) or more commonly watt hours. (1 Wh = 3600 J). The total energy of a cell is sometimes rather confusingly termed the 'watt hour capacity'. As discussed above, the cell voltage, E, deviates progressively from its (maximum) thermodynamic value as the rate of discharge increases. Hence the **energetic efficiency** $\mathscr{E}_P/\mathscr{E}_T$ is a variable quantity, which must be associated with closely defined discharge conditions if it is to be meaningful.

Theoretical and practical energies can also be expressed in terms of the complete discharge of a particular cell:

$$\mathscr{E}'_T = x(nFE_{cell}) \qquad (2.82')$$

and
$$\mathscr{E}'_P = \int_0^{xnF} E.dq \qquad (2.83')$$

where x is again the number of moles of reaction associated with the complete discharge.

The **energy density** (also known as **specific energy**) is the parameter used when assessing relative cell performance. Thus a small battery, weighing 25 g, and capable of delivering 40 kJ or 0.012 kWh at a particular discharge current, would be said to have an energy density of 480 Wh kg^{-1}. Again it is sometimes more useful to consider a volume based specific energy (e.g. Wh dm^{-3}).

Power

The level of discharge current drawn from a cell is determined principally by the external load resistance. The power delivered, P, is given by the product of the current flowing and the associated cell voltage:

$$P = iE \qquad (2.84)$$

The power rating of a battery specifies whether or not it is capable of sustaining a large current drain without undue polarisation. As more and more current is drawn from a cell, the power initially rises; it reaches a maximum and then drops as the cell voltage falls away due to polarisation effects (Fig. 2.21). The maximum power point is best determined experimentally by measuring E as a function of i. In certain circumstances it is possible to calculate the maximum power point – e.g. if electrode polarisation is small and the internal resistance of the cell is known. The **rated power** of a battery is the power delivered (in watts) under stated discharge conditions.

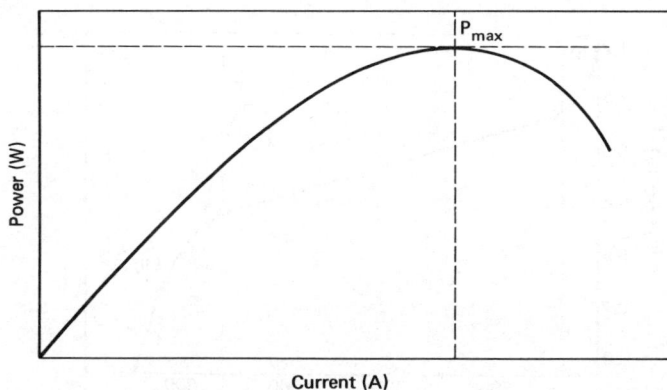

2.21 The power of a typical electrochemical cell as a function of discharge current.

The **rated power density** (or **specific rated power**) is again the most convenient parameter for comparing different battery systems, and is generally quoted in W kg^{-1} or W dm^{-3}. The rated power density and rated energy density are both critical factors when assessing batteries for applications such as motor traction where the battery itself has to be transported. A cell generally has a **maximum permitted continuous power level:** prolonged discharge above this value is liable to cause overheating and a consequent degradation of the cell, as discussed below. On the other hand, cells may be capable of a much higher **rated instantaneous power**. For short discharges the battery does not reach a thermal stationary state and its thermal capacity may be able to accommodate the heat generated if the thermal conductivities of its constituents are sufficiently high. Further, mass transport polarisation effects are of less significance for short discharges. Certain batteries, such as those used for starting large diesel engines, depend specifically on their high instantaneous power capabilities.

As will be seen later, cells employing the same chemical system can be designed either for high power or high capacity.

Polarisation and Discharge Curves

More detailed information on the characteristic behaviour of cells under load and at various stages of discharge can be conveyed by graphs. Plots of cell voltage against current are usually referred to as **polarisation** (or **performance**) **curves**; graphs of cell voltage as a function of the fraction of discharge completed are known as **discharge curves**.

A typical cell polarisation curve (Fig. 2.22) has three regions. Initially there is a rapid fall in cell voltage at low current drain, due to electrode polarisation overvoltage, usually associated to a large extent with one of the two electrode processes. (In a few cases where a passive film on one of the electrodes initially interferes with the discharge, the initial stages may be associated with a rise in cell voltage as the film is destroyed – see Chapter 5.)

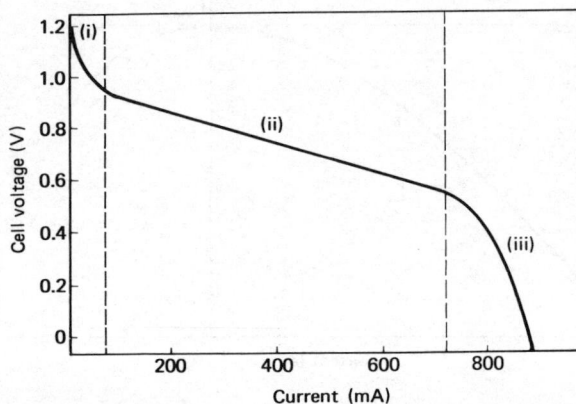

2.22   Typical polarisation curve for an electrochemical cell.

In the second almost linear region, the internal resistance of the cell components causes a further voltage loss (the '*iR* polarisation'). In the final region at relatively high current drain, the *iR* polarisation is combined with further electrode polarisation caused by depletion of electroactive materials at the electrode surfaces. While such curves provide much useful information, they must be interpreted with some care. First, it is often necessary to examine polarisation curves at various stages of discharge since both electrode and electrolyte characteristics can alter considerably as discharge products build up. Second, polarisation curves are usually time dependent, so that the details of how they have been determined must be examined: e.g. has there been a relatively fast sweep of voltage or current through the requisite values, or has a slow, pseudo-steady state approach been taken?

Discharge curves show either the open circuit voltage of a cell or half-cell as a function of the fraction of discharge completed, or, much more commonly, the cell voltage during a deep discharge, usually under a fixed load or at constant current. The abscissa may be calibrated in terms of the quantity of electricity passed (Ah) or as a percentage of the theoretical capacity. A useful method of characterising the discharge (or charge) rates used either in recording discharge curves or for determining practical capacities, is to standardise the current in terms of the nominal cell capacity. A **C-rate** of τ implies that the nominal capacity of the cell (measured in Ah) is delivered in $1/\tau$ hours: e.g. for a 2 Ah cell, discharge at C/5 signifies a current drain of 0.4A.

Discharge curves for typical cathodic half-cells are shown in Fig. 2.23. In curve (a), reactants and products are both in liquid or solid solution, while curve (b) represents the case where they form two distinct solid phases. In Fig. 2.24, a two-stage discharge takes place, where all three electroactive components form separate solid phases and the electrolytic phase remains virtually invariant, e.g.

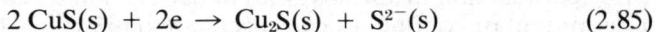

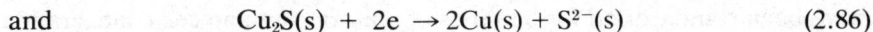

$$2\,CuS(s) \;+\; 2e \;\rightarrow\; Cu_2S(s) \;+\; S^{2-}(s) \tag{2.85}$$

and
$$Cu_2S(s) \;+\; 2e \;\rightarrow 2Cu(s) + S^{2-}(s) \tag{2.86}$$

A number of empirical equations have been suggested (e.g. Shepherd formulae) for certain types of cell which give cell voltage as a function of discharge depth, current, etc., once a number of constants characteristic of the cell discharge have been evaluated.

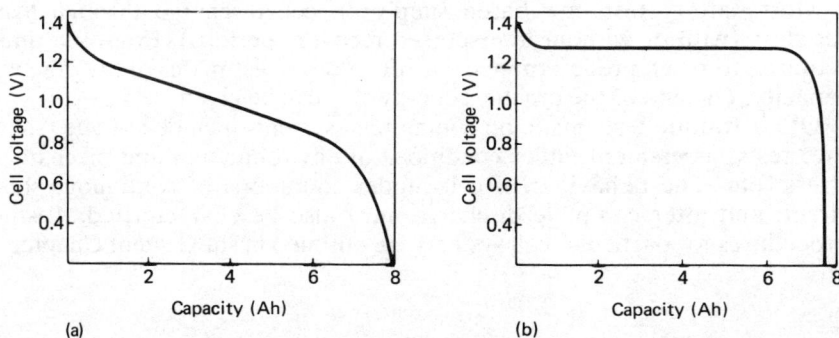

2.23 Discharge curves for typical cathodic half-cells.
(a) reactants and products in solution phase – e.g. Fe^{3+}(aq) and Fe^{2+}(aq);
(b) reactants and products both form solid phases – e.g. Ag_2O(s) and Ag(s).

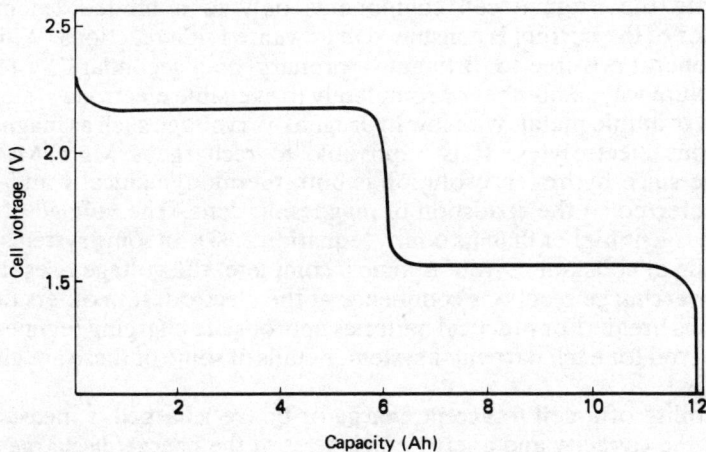

2.24 Two stage discharge curve where all three electroactive components form separate solid phases, as in the cell $Li(s)|LiClO_4$, $PC|CuS(s)$.

Battery Testing and Specifications

The useful life of a practical primary battery is determined principally by the nature of its discharge pattern. Thus the best way of assessing a system for some particular application is to subject it to a discharge which simulates the service conditions. Tests have therefore been developed which recognise the principal function of various types of battery and specify the generation of intermittent or continuous currents of appropriate levels. Such procedures have been standardised for batteries of uniform size and cell configuration by bodies such as the International Electrotechnical Commission (IEC), the American National Standards Institute (ANSI), etc. New test routines are continually being devised to keep pace with new commercial battery developments and applications. Other tests and specifications have been detailed by various military, technical and commercial organisations.

Most battery tests are based simply on cell discharge through fixed resistors (with or without interspersed recovery periods). From the times required to reach predetermined cut-off voltages, estimates of the practical capacity, Q_P, and of the practical energy, $_P$ can be given.

Other routine tests made on commercial systems include storage (shelf-life) tests, assessment under conditions of environmental and mechanical stress, etc. The behaviour of cells under conditions of continuous short circuit and after complete discharge must also be characterised. Testing procedures for particular cell systems are outlined in subsequent Chapters.

Rechargeable systems

Spontaneous cell reactions can often be reversed by applying a potential difference to the cell in such a way as to drive current through it in the opposite direction to the discharge. In some cases the reverse current will regenerate the original cell components only to a limited extent: the remainder of the current is consumed in unwanted side reactions. While it is not in general possible to distinguish 'primary' and 'secondary' systems in any absolute way, a number of completely irreversible electrode systems do exist, for example metals with low hydrogen overvoltage such as magnesium in aqueous electrolytes. It is impossible to recharge a $Mg(s)|Mg^{2+}(aq)$ electrode since hydrogen evolution is both thermodynamically and kinetically preferred to the reduction of magnesium ions. The voltage of a cell under charge is higher than its e.m.f. (equation 2.43). In some systems when recharging at constant current is almost complete, the voltage rises sharply as new **overcharge** reactions commence at the electrodes; in others there is no obvious break. For practical batteries appropriate charging regimes have been derived for each particular system: details of some of these are given in Appendix 1.

The ability of a cell to **accept charge** or be **(re-)charged** is measured in terms of the capacity and energy efficiencies of the **charge/discharge cycle**. The **capacity or ampere-hour efficiency** of a cell cycled under stated conditions of rate and depth is defined as

$$\frac{\int_0^t i_{dis}\, dt}{\int_0^t i_{ch}\, dt}$$

where i_{dis} and i_{ch} refer to the currents flowing during discharge and charge respectively. The inverse of the capacity efficiency is known as the **charge factor**. The charge factor is substantially greater than unity if significant side reactions occur while the cell is being charged. The overall **cycle energy efficiency** is given by

$$\frac{\int_0^t E_{dis}\, i_{dis}\, dt}{\int_0^t E_{ch}\, i_{ch}\, dt}$$

where E_{dis} and E_{ch} are the cell voltages during discharge and charge respectively. It is emphasised that the ability of a cell to accept charge is dependent on the way in which the charging operation is carried out and that these efficiency ratings are not absolute quantities. In practice, cycle efficiencies are usually quoted in terms of particular C-rates for charge and discharge to a fixed depth.

Cycle life is the number of times a cell can undergo a charge/discharge sequence before its performance (as measured by its capacity or energy storage efficiency) has been degraded to below some arbitrary limit. For shallow discharge (say <20%) a much longer cycle life is obtained. Standard tests generally require a discharge to 75% or 80% of rated capacity. **Charge retention** is a measure of whether the percentage of nominal capacity lost as a function of time due to self-discharge reactions occurring within the cell is large or small. Charge retention is much poorer at high temperatures.

In addition to having a high cycle efficiency, a practical rechargeable cell must meet a number of other criteria. In particular it must have a satisfactory rate of charge acceptance. This is governed by the same factors that control the magnitude of normal discharge currents: exchange current, area and nature of electrode surface, ohmic resistance, etc. The latter is of special importance in charging processes since if high charging voltages are applied the problem of overheating and its effect on the chemical and mechanical stability of the system may become critical. Second, if solid phases are formed during recharge, their morphology must be suitable: dimensionally stable, coherent, non-dendritic (especially with metallic phases lest internal short circuits occur), etc.

Because of competing side reactions, the full charge capacity of many secondary battery systems can only be achieved by **overcharging**, i.e. by passing charge in excess of the theoretical capacity during the recharge. In aqueous cells the side reactions associated with the excess charge usually involve electrolysis to form hydrogen or oxygen. The nature of side reaction products can sometimes be controlled by regulating the relative capacity of anode and cathode. If a gas is formed, provision must be made for its venting, or, as in the case of sealed cells, recombination.

Certain problems sometimes arise with rechargeable multicell batteries which are associated with non-uniform capacities of the individual cells. In charging such batteries there must be either overcharge in some cells or undercharge in the whole system. More importantly, on deep discharge, **cell reversal** may take place where the lowest capacity cell suffers a polarity reversal. With some, but not all, systems such an occurrence is catastrophic and a number of electrochemical and electrical reversal protection devices have been incorporated in practical rechargeable multicell assemblies.

Thermal management

'Thermal management' is defined as the method of operation, for a particular engineering design, whereby a battery system is maintained within a specified temperature range while undergoing charge or discharge processes. Cells are generally characterised by a maximum working temperature, above which corrosion and other irreversible destructive processes are unacceptably rapid; cells also have, in practice, a minimum working temperature, below which the electrolyte has too high a resistance, or is liable to undergo a phase change – this is of most importance in 'high temperature' cells (Chapter 6).

The thermal energy generated or absorbed by an electrochemical cell is determined first by the thermodynamic parameters of the cell reaction, and second by the overvoltages and efficiencies of the electrode processes and by the internal resistance of the cell system. While the former are generally relatively simple functions of the state of charge and temperature, the latter are dependent on many variables, including the cell history.

Overheating problems are most common in large batteries (i.e. those with low surface to volume ratios) undergoing high rates of charge or discharge. Some of the thermal energy generated by the cell process may be dissipated as heat to the surroundings* either by natural air convection, possibly augmented by cooling fins, channels or gaps between the cells, etc., or by use of heat exchangers involving forced circulation of gas or liquid coolant or pumping of the electrolyte through an external cooling unit. The remaining energy remains within the cell, resulting in an increase in temperature. The continuous rated power level of a battery is set at such a value that the steady state temperature does not exceed the maximum working temperature.

The thermal management problems of high temperature batteries are more complex. First, a heating system is required to raise the temperature of the battery to the operating value, or to maintain it at this value when no current is being drawn; second, an insulating container must be provided to minimise heat loss and thus raise the overall energy efficiency; and finally, a cooling system may be necessary to prevent overheating during high rate operation.

With the growing interest in large high rate batteries for traction and load levelling applications the development of models for heat transfer analysis has become of increasing importance to design engineers in recent years. Once the thermal properties (heat capacity, thermal conductivities, etc.) are known a number of numerical and analytical models are available for calculating temperature distribution during various operating schedules.

---

* Problems concerning heat dissipation in spacecraft batteries arise as a consequence of the absence of convective heat transfer in a vacuum.

3 Primary aqueous electrolyte cells

3.1 Introduction

The aqueous electrolyte battery group includes some of the oldest but still most commercially important primary and secondary systems. Thus despite considerable advances in the design and manufacture of new types of cell over recent years, the classical Leclanché cell and its variants, based on zinc and manganese dioxide, now more than a century old, is still by far the primary cell in largest production. In the USA alone, total sales exceeded $900 m. in 1980. The strong market position of this cell is due to a combination of a number of factors including low cost of materials and ease of manufacture, together with performance characteristics which are suitable for a wide range of practical applications.

With the continuing development of portable electric and electronic equipment, demand for small primary aqueous electrolyte cells continues to expand and a growth rate in world production in the range of 6–12% per annum has been maintained over the past twenty five years. It is estimated that there is now an annual consumption averaging 8–15 cells per person in the Western developed countries.

Almost all modern practical aqueous primaries are referred to as 'dry cells'. This designation should not be confused with the rather specialised 'solid-state cells' which make use of the recently discovered true solid electrolytes. Rather, the term implies that the aqueous electrolyte phase has been immobilised by the use of gelling agents or by incorporation into microporous separators. Such procedures permit the cells to operate in any orientation and reduce the effects of leakage, should the container become punctured.

The range of commercially available cells is so large that it would not be possible to include here a comprehensive description of their design and performance characteristics. Instead, examples of the more important cells are discussed in detail, together with features of some other systems of interest which have often been developed with particular applications in mind.

3.2 Leclanché cells

The name 'Leclanché cell' is given to the familiar primary system consisting of a zinc anode, manganese dioxide cathode and an electrolyte of ammonium chloride and zinc chloride dissolved in water. The alternative designation 'zinc-carbon cell' is broader and includes the so-called zinc chloride system which, due to a different electrolyte composition, is characterised by a different discharge mechanism. The Leclanché cell may be written as

$$Zn(s)|ZnCl_2(aq),NH_4Cl(aq)|MnO_2(s),C(s)$$

and the OCV* is in the range 1.55–1.74 V. Despite the many advances which have brought this cell to a high degree of reliability and greatly improved performance, the electrochemical system of the modern cell is basically the same as that proposed by George Leclanché in 1866. (Fig. 3.1)

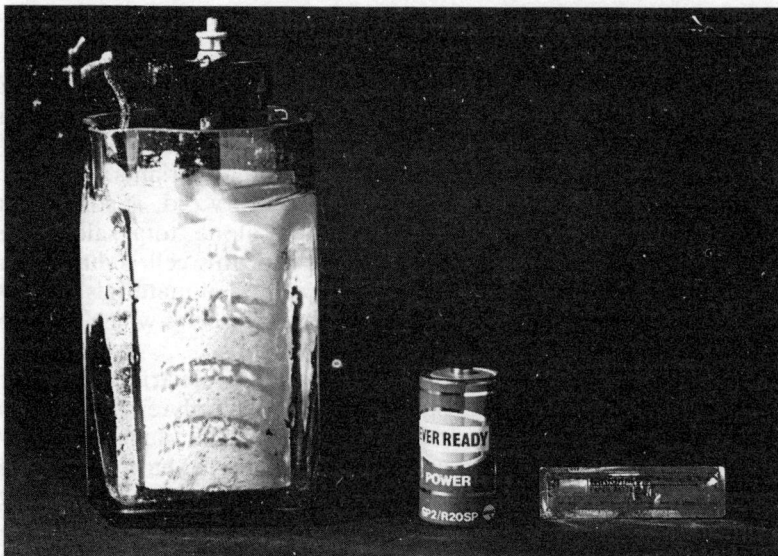

3.1   Leclanché cells, past and present
(i)   Leclanché wet cell;
(ii)   D-size dry cell (Berec);
(iii)   'Paper Battery' (By courtesy of Matsushita Battery Industrial Co.)

The original Leclanché cell was a 'wet cell' assembled in a glass jar. The cathode was a carbon rod or plate immersed in a mixture of manganese dioxide and carbon powder contained in a porous ceramic pot. The anode was a zinc rod which was immersed in the electrolyte, a saturated solution of ammonium chloride (Fig. 3.2). By 1868 more than 20 000 of these cells were in service in European telegraph systems. Twenty years after the original cell had been proposed, Carl Geissner patented the idea of manufacturing the zinc anode in the form of a cup which would act as a container for the electrolyte (which was immobilised with plaster of Paris). Combination of this development of a zinc container with the use of a manganese dioxide/ carbon anode and an electrolyte immobilised using a cereal paste, laid the foundations of modern dry cell technology and led quickly to a massive

* OCV : open circuit voltage – see Glossary

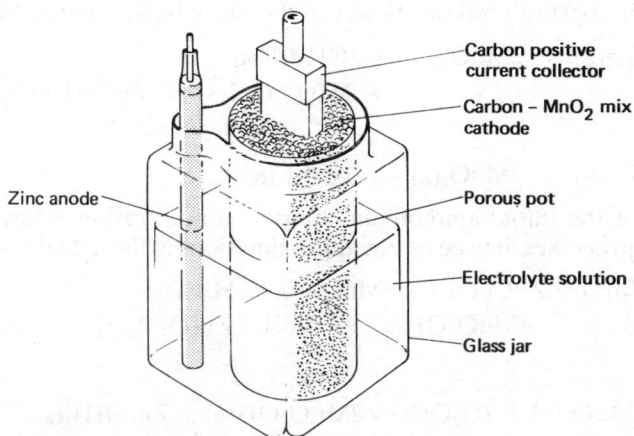

3.2   Schematic view of the original Leclanché wet cell.

production of Leclanché cells which had exceeded an annual production of two million units by the turn of the century.

Discharge mechanisms

The discharge mechanism of the Leclanché cell is complex and not all the details are yet fully understood. The basic process consists of oxidation of zinc at the anode to form zinc ions in solution:

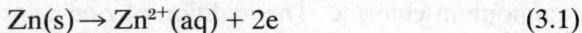

$$Zn(s) \rightarrow Zn^{2+}(aq) + 2e \qquad (3.1)$$

together with a reduction of Mn (IV) to a trivalent state as $MnO.OH(s)$ or $Mn_2O_3.H_2O(s)$, at the cathode, e.g.

$$2MnO_2(s) + 2H_2O + 2e \rightarrow 2MnO.OH(s) + 2OH^-(aq) \qquad (3.2)$$

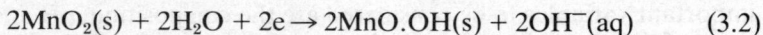

The initial products of the electrode reactions may then undergo further reactions in solution. The prevailing process is:

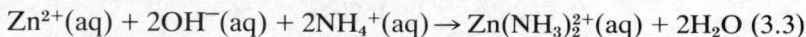

$$Zn^{2+}(aq) + 2OH^-(aq) + 2NH_4^+(aq) \rightarrow Zn(NH_3)_2^{2+}(aq) + 2H_2O \quad (3.3)$$

followed by the formation of the slightly soluble $Zn(NH_3)_2Cl_2$:

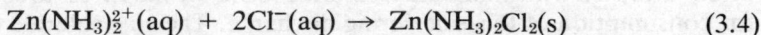

$$Zn(NH_3)_2^{2+}(aq) + 2Cl^-(aq) \rightarrow Zn(NH_3)_2Cl_2(s) \qquad (3.4)$$

For light discharges and with certain oxides an alternative reaction is:

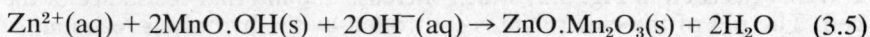

$$Zn^{2+}(aq) + 2MnO.OH(s) + 2OH^-(aq) \rightarrow ZnO.Mn_2O_3(s) + 2H_2O \quad (3.5)$$

At lower NH_4^+ concentrations, the zinc ions precipitate out as one or more oxychloride species: e.g.

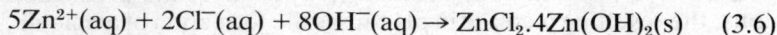

$$5Zn^{2+}(aq) + 2Cl^-(aq) + 8OH^-(aq) \rightarrow ZnCl_2.4Zn(OH)_2(s) \qquad (3.6)$$

or as the hydroxide:

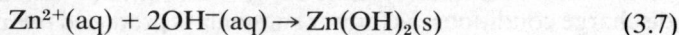

$$Zn^{2+}(aq) + 2OH^-(aq) \rightarrow Zn(OH)_2(s) \qquad (3.7)$$

The principal overall cell reactions can therefore be summarised as

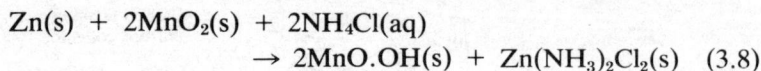

$$Zn(s) + 2MnO_2(s) + 2NH_4Cl(aq)$$
$$\rightarrow 2MnO.OH(s) + Zn(NH_3)_2Cl_2(s) \quad (3.8)$$

and

$$Zn(s) + 2MnO_2(s) \rightarrow ZnO.Mn_2O_3(s) \quad (3.9)$$

However if the initial ammonium chloride concentration is low, then the following processes may be better approximations to the actual cell reaction:

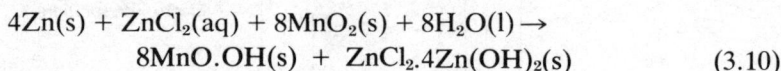

$$4Zn(s) + ZnCl_2(aq) + 8MnO_2(s) + 8H_2O(l) \rightarrow$$
$$8MnO.OH(s) + ZnCl_2.4Zn(OH)_2(s) \quad (3.10)$$

and

$$Zn(s) + 2MnO_2(s) + 2H_2O(l) \rightarrow 2MnO.OH(s) + Zn(OH)_2(s) \quad (3.11)$$

Calculations of the e.m.f. of cells based on these reactions provide values within the wide range of 1.5–1.7 V, characteristic of undischarged cells formed with different samples of manganese dioxide and electrolyte pH. However during the discharge of practical cells, inhomogeneities in the solution and cathode phases may produce a much more complicated reaction sequence, as will be discussed below.

Electrolyte

The electrolyte in the original Leclanché cell was a saturated solution of ammonium chloride. The addition of zinc chloride was soon found to be beneficial to cell performance, and since Geissner's time the electrolyte has contained both salts. The ternary phase diagram for the system $ZnCl_2$–NH_4Cl–H_2O at room temperature is shown schematically in Fig. 3.3. The important variables to be considered are the composition, pH and conductivity of the electrolyte. A typical composition for an undischarged cell is:

NH_4Cl	$ZnCl_2$	H_2O
28%	16%	56%

although there are variations depending on manufacturer. The solution is often saturated with ammonium chloride to enable the excess to compensate for consumption of this salt during discharge. The zinc chloride content is limited by the formation of the solid phases $ZnCl_2.3NH_4Cl$ and $ZnCl_2.2NH_4Cl$ (as seen in Fig. 3.3) which increase the internal resistance of the cell. The pH of the electrolyte is controlled by the hydrolysis reactions of both the ammonium and the zinc ions, and increases with increasing zinc or ammonium ion concentration.

Since the basic cathode reaction (3.2) involves the production of OH^- ions, the resulting increase in pH causes the formation of ammonia, which in turn complexes with zinc ions. Initially a precipitate of slightly soluble diamminozinc chloride, $Zn(NH_3)_2Cl_2$ is formed, but according to the discharge conditions, as further ammonia is produced this may be converted to the more soluble tetramminozinc chloride, $Zn(NH_3)_4Cl_2$:

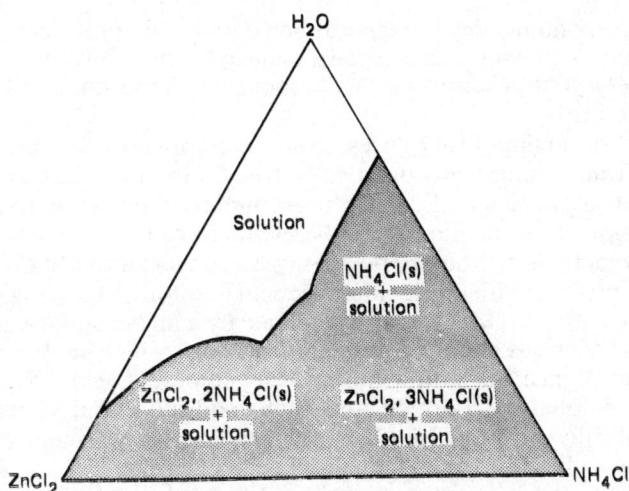

3.3 Schematic ternary phase diagram for the system ZnCl$_2$–NH$_4$Cl–H$_2$O at room temperature.

$$Zn^{2+}(aq) + 2Cl^-(aq) \xrightarrow[\text{Slightly soluble}]{2NH_3} Zn(NH_3)_2Cl_2 \xrightarrow{2NH_3}$$

$$Zn(NH_3)_4^{2+}(aq) + 2Cl^-(aq)$$

$$(3.12)$$

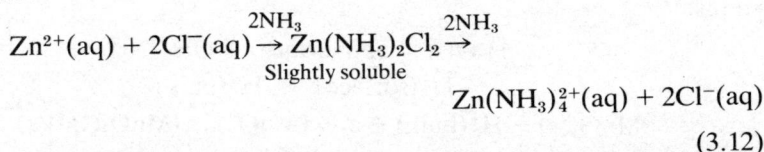

The formation of these complexes serves to buffer the rise in pH due to the cathodic discharge process. Precipitation of Zn(NH$_3$)$_2$Cl$_2$ has the effect of increasing cell resistance.

In the anodic region, the concentration of Zn^{2+} ions increases as discharge proceeds, leading to a decrease in pH due to hydrolytic reactions of the type:

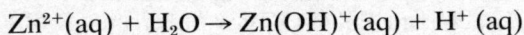

$$Zn^{2+}(aq) + H_2O \rightarrow Zn(OH)^+(aq) + H^+(aq)$$

A pH gradient is thus established during cell discharge.

Cathode

The cathode 'mix' consists basically of manganese dioxide to which carbon, in finely divided form (e.g. as acetylene black), has been added in order to provide an adequate electronic conductivity. The current collector is generally also made of carbon, in the form of either a rod or a thin sheet. The active material is the MnO$_2$ and it was realised from early in the development of the Leclanché cell that use of naturally occurring MnO$_2$ from different sources produced striking variations in cell performance. Even now, when synthetic MnO$_2$ is used in almost all high performance batteries, slight changes in the method of preparation can have significant effects. MnO$_2$ has numerous allotropic forms, with subtle crystal structure modifications and a wide variety of surface types. So far it has not proved possible to relate in any exact way the crystal structure, surface properties, etc. of a particular sample of MnO$_2$

with its corresponding electrochemical behaviour. The only decisive test of the effectiveness of MnO_2 as a battery material is its behaviour in the cell. Much has been written about the 'art' rather than the 'science' of Leclanché cell manufacture!

Naturally occurring MnO_2 ores from a number of sources (and in particular from Ghana) are practically free from significant amounts of heavy metal impurities and can be used directly after washing processes without chemical purification. Improvements in cell performance may be obtained by 'activating' the ore by heating to convert it to Mn_2O_3 and then treating the product with dilute sulphuric acid to reform MnO_2 together with $MnSO_4$. The activated MnO_2 is characterised by a higher surface area.

'Synthetic' MnO_2 is obtained from natural MnO_2 which has been reduced to MnO and thence put into solution using sulphuric acid. The resulting manganous sulphate is purified and then oxidised either chemically or electrochemically to yield MnO_2 with high surface area and consistent properties, which is used in high quality Leclanché cells.

The overall cathodic reaction (equation 3.2) in which $MnO_2(s)$ is reduced to $MnO.OH(s)$ is now known to involve a solid state diffusion process in which protons are transported from the surface to the interior of the MnO_2 grains:

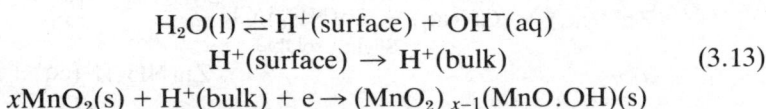

$$H_2O(l) \rightleftharpoons H^+(\text{surface}) + OH^-(aq)$$
$$H^+(\text{surface}) \rightarrow H^+(\text{bulk}) \qquad (3.13)$$
$$xMnO_2(s) + H^+(\text{bulk}) + e \rightarrow (MnO_2)_{x-1}(MnO.OH)(s)$$

Initially the cathodic product contains a variable amount of trivalent manganese in a homogeneous phase; at higher levels of discharge a new crystalline modification composed solely of $MnO.OH$ is also formed. As shown in equation 3.5, these products may react further with zinc ions to form mixed oxides. The composition of the cathodic discharge product is thus seen to vary in a very complex manner as discharge proceeds, and this is responsible for the characteristic fall in OCV with percentage of service life expended (see, for example, Fig. 3.12 and 3.15).

In addition, the slow diffusion of protons within the solid MnO_2 results in serious polarisation of the cell, especially at high currents (Fig. 3.4) (a)). If the cell is allowed to 'rest' for a period while no current is drawn, the cell voltage slowly recovers, as proton diffusion dissipates the high Mn (III) concentration near the surface of the MnO_2 grains and the composition of the solid returns to uniformity. (Fig. 3.4 (b)). (The recuperation of the cell voltage is also aided by the dispersion of the pH gradient across the cell which also depresses the OCV.)

Because of the nature of the cathodic discharge, it can be appreciated that the Leclanché cell must have severe limitations in high current operations. On the other hand in applications requiring intermittent use, the full beneficial effects of the recuperation process may be experienced. Fig. 3.5 shows the intermittent discharge curve of a C-size cell subjected to a standard 'Heavy Industrial Flashlight Test' (4 Ω load, 4 minute discharge every 15 minutes over an eight hour period, repeated daily): the voltage recovery during the rest periods is clearly seen. The rate of the recuperation process is affected by the nature of the MnO_2 used.

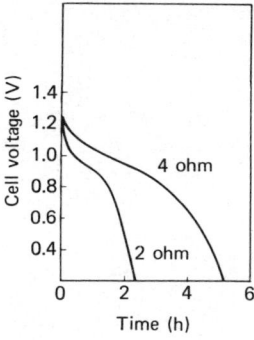

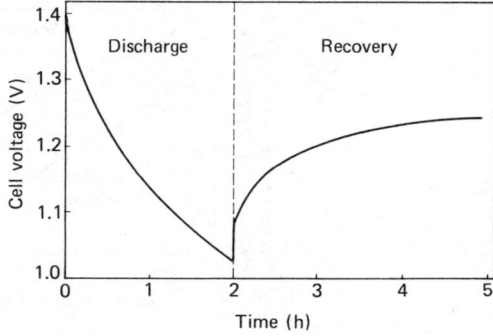

3.4(a) Continuous discharge curves of a D-size Leclanché cell through 2 ohm and 4 ohm loads.

3.4(b) Recovery of cell voltage after discharge through a 4 ohm load for two hours.

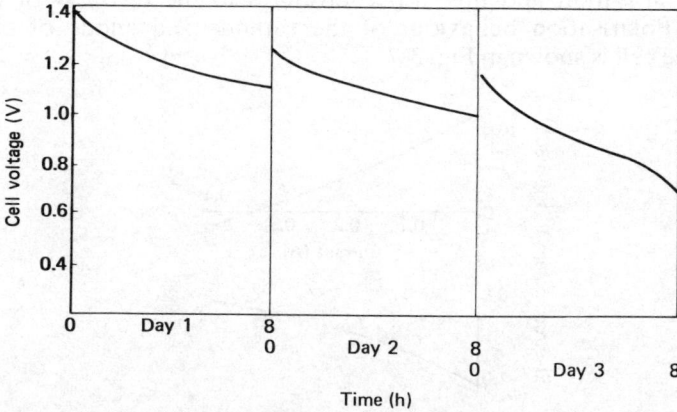

3.5 Intermittent discharge curve of a C-size Leclanché cell through a 4 ohm load using an HIF (Heavy Industrial Flashlight) test schedule: 4 minute discharge every 15 minutes over an eight hour period, repeated daily.

Anode

In modern cells the anode consists of a zinc alloy sheet, containing small quantities of Pb and Cd to give satisfactory mechanical properties for drawing, extrusion, etc., often lightly amalgamated with mercury. The potential of the zinc electrode with respect to a saturated calomel electrode is shown in Fig. 3.6 as a function of $ZnCl_2$ concentration. In the presence of NH_4Cl the potential is shifted to more negative values due to the formation of chlorozinc complexes (and amminozinc complexes at higher pH) with the consequent reduction in the activity of free zinc ions.

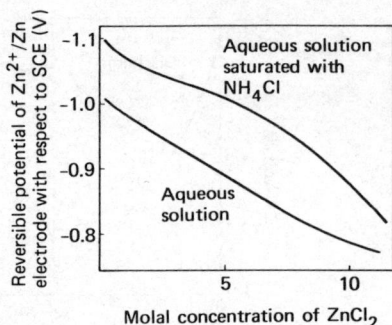

3.6   Reversible potential of a zinc electrode with respect to a saturated calomel electrode (SCE) as a function of zinc chloride concentration.

Electrode polarisation of the anode is less severe than that for the cathode, and is mainly concentration polarisation resulting from accumulation of zinc chloride near the electrode surface. During rest periods after discharge, diffusion of $ZnCl_2$ into the bulk of the electrolyte reduces the anode polarisation and this also contributes to the recovery of the cell voltage. Polarisation behaviour of the cathode and anode of a D-size Leclanché cell is shown in Fig. 3.7.

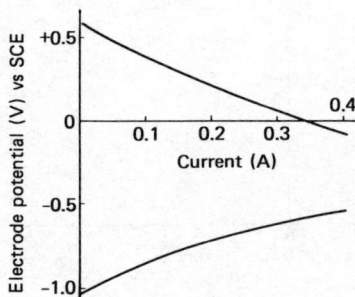

3.7   Polarisation curves for anode and cathode of a D-size Leclanché cell: current increased by steps of 25 mA once per minute.

Shelf reactions

Shelf reactions are defined as chemical processes occurring in batteries during storage before they have been used. Similar processes also occur during rest periods between discharges, but where changes in electrolyte composition have occurred as a result of discharge (as happens in the case of the Leclanché cell) the nature and rates of these reactions may be greatly changed. The most significant shelf reactions which lead to the deterioration of Leclanché cells are corrosion processes at the anode. Three main mechanisms must be considered. First, if oxygen can reach the zinc/solution interface, the reaction

$$Zn(s) + 2NH_4^+(aq) + 2Cl^-(aq) + \tfrac{1}{2}O_2(g) \rightarrow$$
$$Zn(NH_3)_2Cl_2(s) + H_2O(l) \qquad (3.14)$$

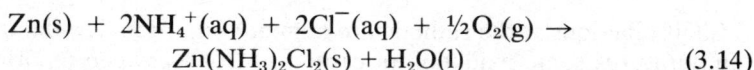

becomes possible. Effective sealing of the cell is therefore necessary, both to exclude oxygen and to prevent evaporation of water from the electrolyte. In the absence of oxygen, a similar corrosion reaction takes place with the evolution of hydrogen – despite the reasonably high hydrogen overvoltage of zinc:

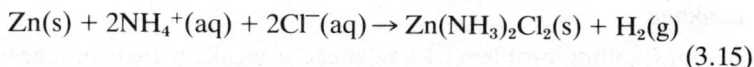

$$Zn(s) + 2NH_4^+(aq) + 2Cl^-(aq) \rightarrow Zn(NH_3)_2Cl_2(s) + H_2(g)$$
$$(3.15)$$

This reaction is greatly enhanced if the third corrosion process is widespread, namely displacement reactions which occur when electropositive impurities such as Ni^{2+}, Fe^{2+}, Cu^{2+} etc. are present in the electrolyte, e.g.:

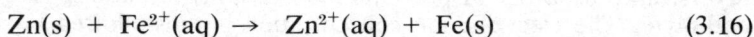

$$Zn(s) + Fe^{2+}(aq) \rightarrow Zn^{2+}(aq) + Fe(s) \qquad (3.16)$$

Such displacements result in the formation of local couples where hydrogen evolution and zinc dissolution can occur at a greatly accelerated rate (Fig. 3.8). (Where the concentration of such impurities is very high, displacement reactions have been known to produce such extensive dendritic growths that the cells have become internally short-circuited.)

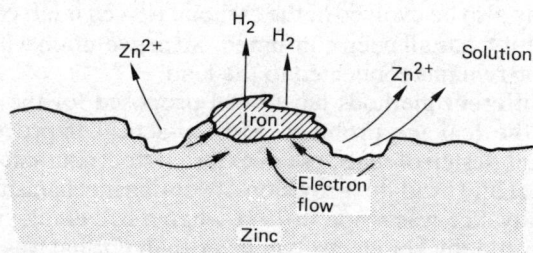

3.8 Mechanism of zinc dissolution by the formation of a local corrosion couple.

The corrosion reactions are strongly temperature dependent and are also affected by the detailed morphology of the zinc and the composition of the electrolyte. Inhibition of corrosion may be achieved by a number of techniques. If a small quantity of a soluble mercury salt is added to the electrolyte, the zinc surface becomes amalgamated. This increases the hydrogen overvoltage of zinc and also dissolves small quantities of electropositive metals, thus preventing the formation of local couples. Amalgamation also influences the morphology of the zinc, producing a smoother surface and reducing localised corrosion. While the addition of a mercury salt has been widely adopted by manufacturers of commercial cells, different companies employ a variety of additional corrosion inhibitors ranging from surface active agents, chelating agents to chromates and dichromates. Since the main source of heavy metal impurities is the MnO_2 in the cathode, careful selection or purification of this material is essential.

Shelf reactions at the cathode are of minor importance, although at high temperatures some oxidation of carbon to CO_2 has been reported. Some MnO_2 may also be lost due to reaction with the gelling agents in the electrolyte.

Water loss from the electrolyte by evaporation or reaction to form hydrates may also take place.

Leakage

A longstanding problem of Leclanché dry cells is their propensity to leak after heavy discharge. The origin of this phenomenon may be traced to the formation of insoluble complexes of zinc (involving chloride, hydroxide and ammonia) in the region between anode and cathode. The establishment of an insoluble layer drastically reduces diffusion between the two electrode layers, and in particular prevents the transport of basic species from the catholyte. The free zinc ion concentration is thus allowed to rise in the anolyte. This in turn causes the pH to fall rapidly, and the increasingly acid solution has a greater tendency to corrode the zinc, with consequent evolution of hydrogen. At the same time, the low pH hydrolyses the starch commonly used to immobilise the electrolyte, and produces a highly viscous phase which tends to trap the hydrogen. The pressure thus developed leads to electrolyte being forced past seals, etc. to the exterior of the cell.

Hydrogen may also be evolved at the cathode if a cell is left connected to a load after the MnO_2 has all been consumed. Such evolution will continue so long as some zinc remains connected to the load.

Very many different methods have been proposed for the reduction or elimination of the leakage problem. Most practical improvements have arisen from better design of seals, use of external steel containers with 'dead volume' to collect any exudate, provision of venting mechanisms, fuses, etc. An important advance was made in 1971 when a cell employing a smaller cathode volume and thicker electrolyte layer than normal was patented. In this cell the complexed zinc precipitate was less coherent and did not interfere significantly with transport between the anode and cathode regions. No leakage was reported even when such cells were short-circuited for 24 hours.

Typical power sources

Leclanché cells are manufactured in a range of sizes, from 11.3 mm diameter × 3.3 mm high button cells to 66.7 mm diameter × 166 mm high alarm cells. Combination of cells in series and parallel arrangements gives a wide variety of batteries with different capacities and voltages ranging from 1.5 V to 510 V. The basic unit is generally constructed in one of two forms, namely the cylindrical cell, used alone or in groups, or the flat cell, used only in the production of multi-cell batteries.

The cross-sections of two typical cylindrical cells are shown in Fig. 3.9. In Fig. 3.9(a) a D-size unit used for flashlights and similar applications is shown; a large capacity alarm cell is shown in Fig. 3.9(b). The electrolyte or

One piece metal cover (+)

Electrode (carbon)

Asphalt seal

Jacket
(polyethylene coated
kraft and polyester
film label)

Paste
(flour, starch,
ammonium chloride
zinc chloride)

Vent washer
(paperboard)

Wax ring seal

Support washer
(Polyethylene coated
paperboard)

Mix
(manganese
dioxide, etc.)

Can (zinc)

Cup (kraft paper)

Bottom (–)
(tin plated steel)

Star bottom
(paperboard)

(a)

Postive terminal binding post

Negative terminal
binding post

Inner
seal asphalt

Seal
support washer

Paste coated
pulpboard
separator

Zinc can:
outside surface
asphalt coated

Cover
(Plastic coated
insulation board)

Expansion
chambers

Carbon electrode

Depolarizing mix

Chipboard
jacket

(b)

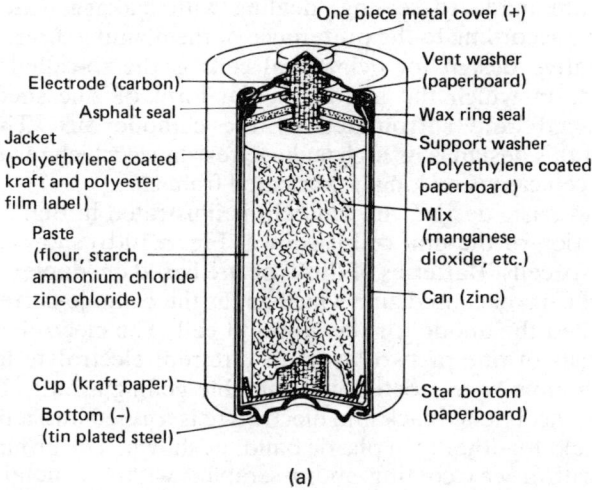

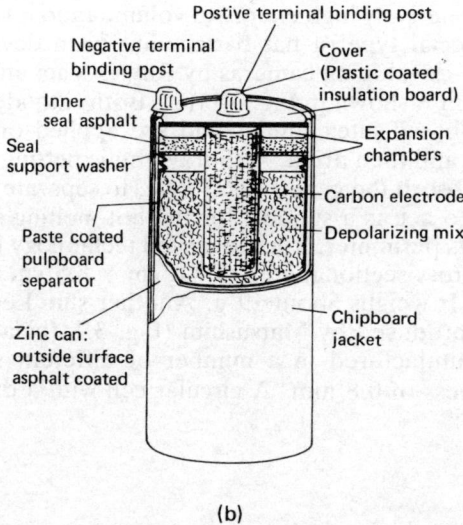

3.9(a) Cross-section of a standard D-size Leclanché cell (By courtesy of Union Carbide)

3.9(b) Cross-section of a high capacity Leclanché alarm cell (By courtesy of Union Carbide)

'paste' separator in Fig. 3.9(a) is a relatively thin layer of electrolyte solution immobilised in a gel or microporous separator. Different manufacturers favour different forms of separator. These range from gelled electrolytes of varying thickness, using either natural cereals or synthetic polyvinyl or cellulose alkyl ethers, pulp board liners soaked in electrolyte solution (as in Fig. 3.9(b)) and special papers coated on either side with films of gelling agent particularly suitable for anolyte or catholyte. Similarly, different approaches are made to sealing, dealing with leakage, external jacket material, etc., according to the preference of the manufacturer.

An alternative design for cylindrical cells is the so-called 'inside-out' configuration, in which the anode, in the form of zinc sheet vanes, is centrally placed, and surrounded by the cathode mix. The principal advantage of this design is its high leakage resistance which results from the fact that the cell case is no longer fabricated from zinc.

The second basic design, the flat cell, is illustrated in Fig. 3.10. In Fig. 3.10(a) a section of the unit cell is shown; Fig. 3.10(b) shows a 9 V series assembly of six cells. Batteries of this type are based on 'duplex' or 'bipolar' electrodes of carbon coated zinc which act as the cathode current collector for one cell and the anode for the adjacent cell. The electrolyte/separator usually consists of one or two layers of different electrolyte impregnated papers which have been treated with suitable gelling agents. The cathode mix of MnO_2, acetylene black and electrolyte is formed into a flat cake and each cell is held together by a plastic band, as shown. The group of six cells are sealed with a wax coating and assembled within a metal jacket. No provision of expansion volume is found necessary with this form of construction, and a relatively high energy to volume ratio is realised.

Recently a special type of flat battery has been developed for use in automatic or self-developing cameras by Ray-O-Vac: an exploded view of this battery system is shown in Fig. 3.11(a). Both cathode and anode mixes contain a water-based latex binder and are applied or 'painted' onto a conductive sheet and then dried. A thin synthetic netting, just over 0.1 mm thick is used to absorb the gelled electrolyte, to separate successive duplex electrodes, and to act as a support for the hot-melting sealing compound applied around its perimeter. This advanced technology battery containing four cells has a cross-sectional area of 7.0 cm × 8.5 cm and a thickness of less than 3 mm. It weighs about 19 g. Another slim Leclanché cell is the 'Paper Battery' produced by Matsushita (Fig. 3.11(b) and Fig. 3.1). This 1.5 V cell is manufactured in a number of different shapes and has a maximum thickness of 0.8 mm. A circular cell with a diameter of 38 mm weighs 1.5 g.

Performance

The practical capacity of a Leclanché cell does not have a fixed value since it varies according to the pattern and conditions of discharge – to a much greater extent than for most other cells. Different forms of this type of cell are designed for particular purposes, and as pointed out in Chapter 2, test procedures imitate as far as possible the discharge pattern of the application. Thus in comparing performance of Leclanché cells it is important always to compare like with like.

The factors which affect the electrical output of a Leclanché cell may be divided into two groups: (i) cell dependent – e.g. construction, composition and size, and (ii) user dependent – e.g. discharge rate, cut-off voltage, operating schedule, operating temperature, storage conditions, etc.

The construction of a cell involves a number of variables such as the relative amount and nature of the cathodic mix, type of separator used, etc.

Plastic cell container

Cathodic mix

Separator

Zinc

Electrolyte impregnated paper

Carbon coating

(a)

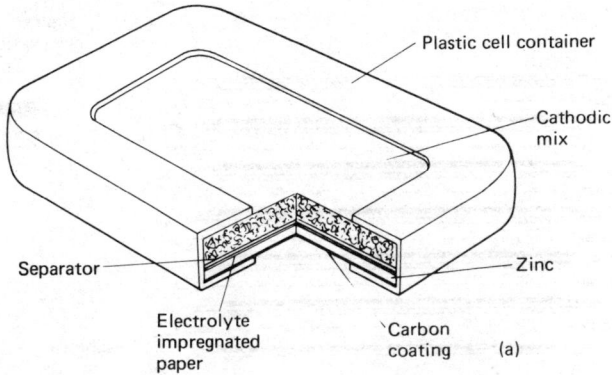

Top plate This plastic plate carries the miniature snap fastener connectors and closes the top of the battery.

Metal Jacket Is crimped on to the outside of the battery and carries the printed design. This jacket helps to resist bulging, breakage and leakage and holds all components firmly together.

Wax coating This seals any capillary passages between cells and the atmosphere; so preventing the loss of moisture.

Plastic cell container This plastic band holds together all the elements of a single cell.

Cathode mix This is a flat cake containing manganese dioxide as the depolariser material and carbon to render it conductive.

Separator Between the mix cake and zinc electrode.

Electrolyte impregnated paper Contains the electrolyte and is an additional separator between cake and zinc.

Carbon coated zinc electrode Known as a Duplex Electrode, it is a zinc plate to which is adhered a thin layer of highly conductive carbon which is impervious to electrolyte.

PVC Covered wire Soldered to the negative zinc plate at the base of the stack and is connected to the negative socket at the other end.

Bottom plate This plastic plate closes the bottom of the battery.

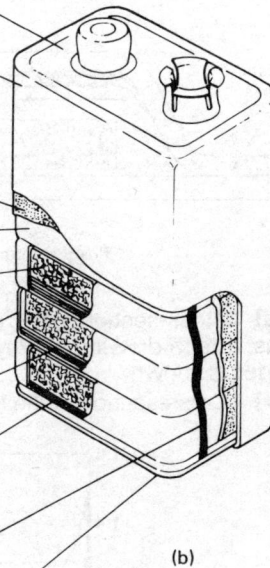

(b)

3.10(a) Cross-section of a flat Leclanché cell.
3.10(b) 9 volt battery of six flat Leclanché cells. (By courtesy of Berec)

Fig. 3.12 illustrates the difference in service life between a standard low power cell and a cell which has been designed for high power applications when both are subjected to a heavy discharge (30 minutes once a day through a 2 Ω load). For a similar construction and composition, and a fixed duty schedule, higher capacities are obviously obtained by using cells of larger size.

The effect of discharge rate is shown in Fig. 3.13 and 3.14 where closed

(a)

(b)

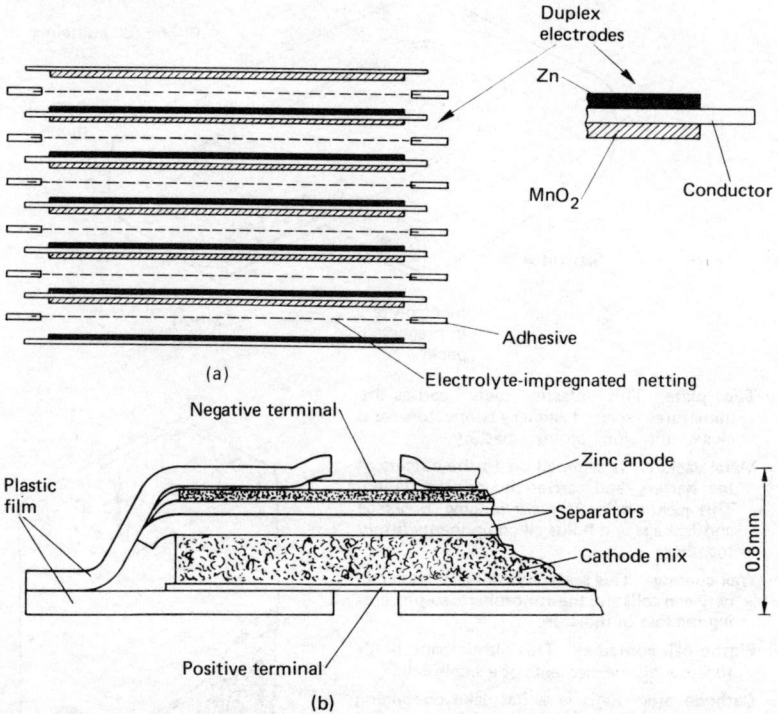

3.11(a)   Cross-section of 3 mm thick Leclanché battery used in automatic cameras, after a drawing by Ray-o-Vac. A detail of the construction of the duplex electrodes is shown.

3.11(b)   Cross-section of the Matsushita 'paper battery'.

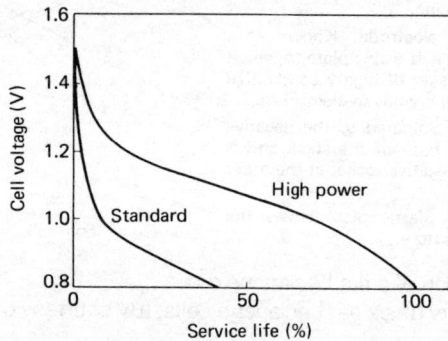

3.12   Difference in service life of standard and 'high power' Leclanché D-size cells discharged for 30 minutes per day through 2 ohm loads.

circuit voltage characteristics and service life of D-size cells are shown as a function of current drain. The fact that service life increases as current density decreases suggests that for this type of system it is always best to use

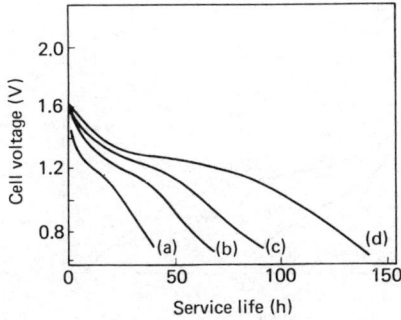

3.13 Effect of discharge rate on service life of D-size Leclanché cells discharged at 2 hours per day
- (a) initial drain : 150 mA
- (b) initial drain : 100 mA
- (c) initial drain : 75 mA
- (d) initial drain : 50 mA

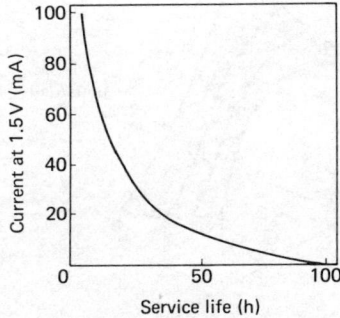

3.14 Effect on service life of initial current drain. D-size standard Leclanché cell discharged at 4 hours per day or to 0.9 V cut-off.

as large a cell as possible. It has been estimated that over a wide range of current densities, the service life is tripled by halving the current density. The service schedule is of critical importance in relation to the recuperation reaction. Unless the current drain is very low, Leclanché cells give a much better performance when used on an intermittent basis. The effect of rest periods was shown earlier in Fig. 3.5. In Fig. 3.15 the effect on the service life of D-size cells of two different operating schedules is illustrated. A schematic three-dimensional representation of capacity as a function of current drain and operating programme based on Union Carbide technical literature is given in Fig. 3.16.

Other factors which affect the electrical output are the operating temperature and the storage conditions. Cells are generally tested at 21 °C (70 °F). Higher temperatures increase the energy output but reduce the shelf-life. The consequences of altering the operating temperature on short and long term discharge are shown in Fig. 3.17. For long term discharge (six months)

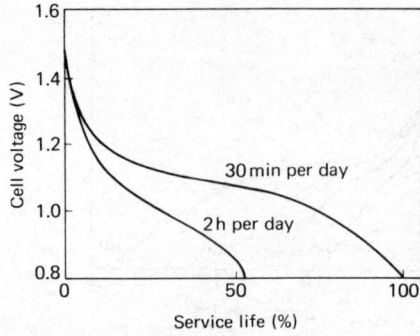

3.15 Effect of operating schedule on service life of a standard D-size Leclanché cell discharged through a 5 ohm load.

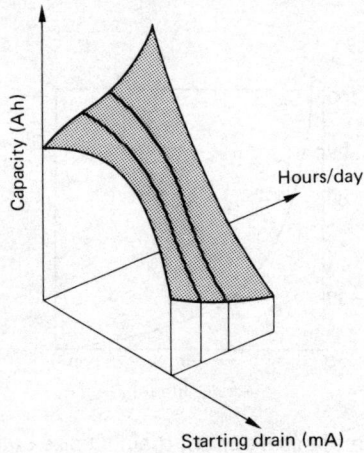

3.16 Capacity of D-size standard Leclanché cells as a function of duty cycle and initial current drain. (By courtesy of Union Carbide)

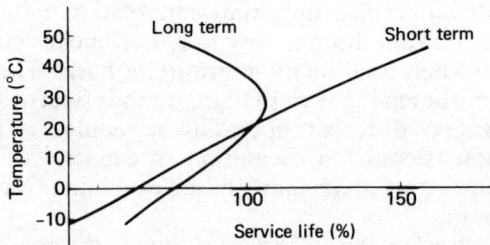

3.17 Effect of temperature on the service life of a 9 V Leclanché flat battery subjected to short term (≈1 month) or long term (≈6 month) discharge (By courtesy of Berec).

at temperatures over 30 °C, the detrimental effects of accelerated shelf reactions are dominant. In Table 3.1 the normalised capacity of D-size cells, discharged continuously through a load resistance of 2.25 Ω to a cut-off voltage of 0.9 V, is given as a function of temperature. As is clearly seen, the low temperature behaviour of normal Leclanché cells is very poor. Special Leclanché cells with altered electrolytes have been used in the past for low temperature applications, but in current practice such cells are replaced by lithium, zinc chloride or zinc/alkaline MnO_2 cells which have greatly superior low temperature behaviour.

Advances in Leclanché dry cell performance have been continuous since the time of Geissner. Between 1900 and 1960 the specific capacity of cells doubled roughly every twenty years. Between 1960 and 1980 a further increase in capacity of over 50% has been achieved. Despite the rapid evolution of new forms of primary battery in recent years, the hold of the Leclanché cell on the market remains very strong for those applications for which its relatively poor discharge characteristics are adequate. The reasons for this are not hard to find: the materials used in its construction are readily available and relatively cheap, while fabrication processes are comparatively straightforward.

Table 3.1

Effect of temperature on the capacity of D-size Leclanché and $ZnCl_2$ cells when discharged continuously through 2.25 Ω to a cut-off voltage of 0.9 V.

Temperature	Normalised capacity	
°C	Leclanché	$ZnCl_2$
37.8	1.40	1.15
26.7	1.10	1.05
21.1	1.00	1.00
15.6	0.90	0.95
4.4	0.70	0.85
− 6.7	0.45	0.70
−17.8	0.25	0.45

3.3 Zinc chloride cells

The so-called 'zinc chloride cells' are basically Leclanché cells in which the ammonium chloride has been completely replaced by $ZnCl_2$. The resulting cells have a better service capacity at high current drain, at low temperatures, and on continuous discharge.

The cell may be written as

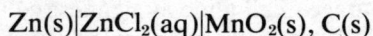

$$Zn(s)|ZnCl_2(aq)|MnO_2(s), C(s)$$

and the OCV is again about 1.5V. The accepted overall cell reaction is

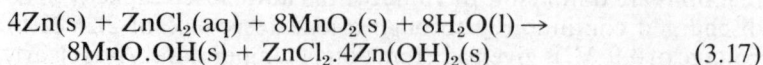

$$4Zn(s) + ZnCl_2(aq) + 8MnO_2(s) + 8H_2O(l) \rightarrow$$
$$8MnO.OH(s) + ZnCl_2.4Zn(OH)_2(s) \qquad (3.17)$$

It should be noted that water is consumed in the cell reaction so that there is a tendency for the cell to dry out during discharge. The anode and cathode are similar to those in the Leclanché cell although higher quality MnO_2 and a higher percentage of acetylene black are normally used. The advantages of this system arise from the properties of the electrolyte where higher rates of diffusion are possible, since there is less tendency for the electrolyte layers near the electrode surfaces to be blocked by insoluble products. The range of operating temperatures is much wider (see Table 3.1) so that this system can form a satisfactory low temperature power supply. Further, higher current densities may be obtained without unacceptable polarisation: for this reason some manufacturers refer to $ZnCl_2$ cells as 'Heavy Duty' batteries.

Fig. 3.18 shows a section of a commercial D-size $ZnCl_2$ cell. A more sophisticated sealing system is generally employed since prevention of leakage is more important with the more acid electrolyte. $ZnCl_2$ cells always use a thin treated paper separator.

The improved performance of the $ZnCl_2$ cell is offset by the higher cost incurred through using a better quality cathode mix and more complex fabrication due to the requirement of more reliable seals.

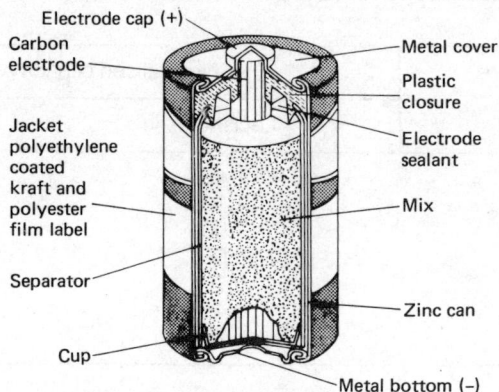

3.18 Cross-section of a D-size zinc chloride cell. (By courtesy of Union Carbide)

3.4 Alkaline manganese cells

This type of cell is another variant on the basic Leclanché cell. In this case the electrolyte is a concentrated aqueous solution of potassium hydroxide (about 30%), partly converted to potassium zincate by the addition of zinc oxide. The main advantage of alkaline manganese cells over Leclanché cells is their relatively constant capacity over a wide range of current drains and under severe service schedule conditions. Another feature of this system is that it can be the basis of a secondary battery system. The cell reaction may be written formally as:

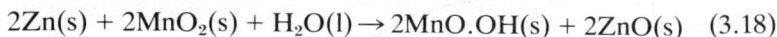

$$2Zn(s) + 2MnO_2(s) + H_2O(l) \rightarrow 2MnO.OH(s) + 2ZnO(s) \quad (3.18)$$

but is in practice much more complex than this due to further reduction of the manganese, as discussed below, and the formation of various soluble zincate species. The OCV is 1.55 V at room temperature. A wet cell based on this system was reported in 1882, but the first commercial dry cell was not available until 1949. Batteries having a wide variety of sizes and voltages are now in use in numerous heavy duty applications.

Electrolyte

A variable quantity of ZnO is added to the concentrated KOH solution, depending on the system characteristics required. The electrolyte is immobilised using carboxymethylcellulose and a non-woven fabric separator made of natural or synthetic fibres resistant to the high pH is placed between the electrodes.

Cathode

The cathode mix is a compressed mixture of electrolytic MnO_2 and graphite in a ratio of 4–5:1, wetted with electrolyte. The cathode current collector is generally the external steel can. Reduction of MnO_2 in alkaline conditions is a complex process and follows a number of steps which can be written formally as

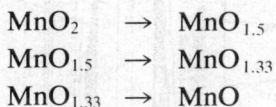

$$MnO_2 \rightarrow MnO_{1.5}$$
$$MnO_{1.5} \rightarrow MnO_{1.33}$$
$$MnO_{1.33} \rightarrow MnO$$

The last two stages are only possible at very low current drain. A schematic diagram of cell voltage as a function of the degree of reduction of MnO_2 is shown in Fig. 3.19. Provided that the reduction does not exceed a level equivalent to $MnO_{1.33}$ the reaction can be reversed and the cathode recharged. In practice this means limiting the discharge at 0.9 V. While the main discussion of secondary aqueous systems follows in the next chapter, the secondary alkaline manganese cell is so similar to the primary cell that it is more convenient to consider it here.

Anode

The anode is a hollow cylinder of powdered amalgamated zinc set in a carboxymethylcellulose gel. The current collector is usually made of brass and the interior space is filled with immobilised KOH solution.

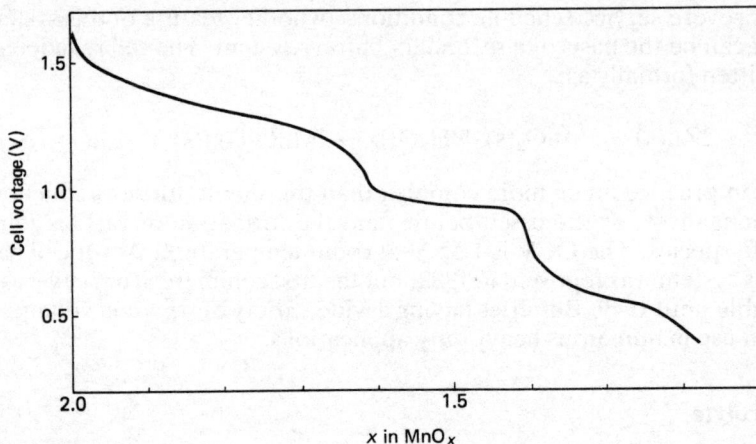

3.19   Cell voltage of an alkaline manganese cell as a function of the degree of reduction of MnO_2.

Typical power sources

A cutaway section of a cylindrical primary cell is shown in Fig. 3.20. The steel construction and intricately engineered sealing assembly reduces the risk of leakage of the highly caustic electrolyte or of release of internal

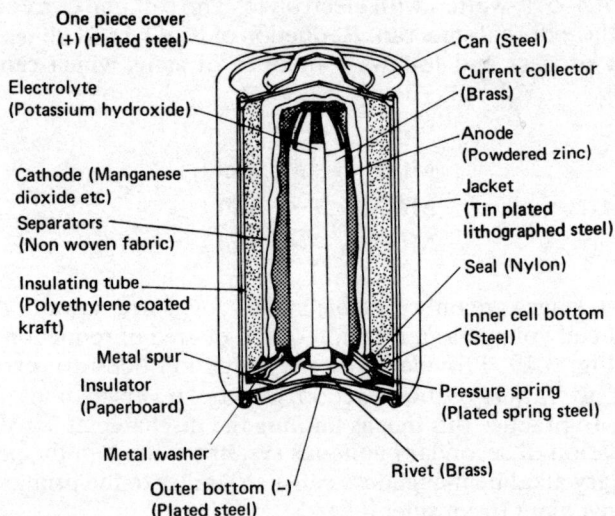

3.20   Cross-section of a D-size alkaline manganese primary cell. (By courtesy of Union Carbide)

pressure. Alkaline manganese secondary cells are also sealed units: a section of a D-size cell is shown in Fig. 3.21. The most important modification is the thicker separator required to prevent short circuits due to dendritic growth

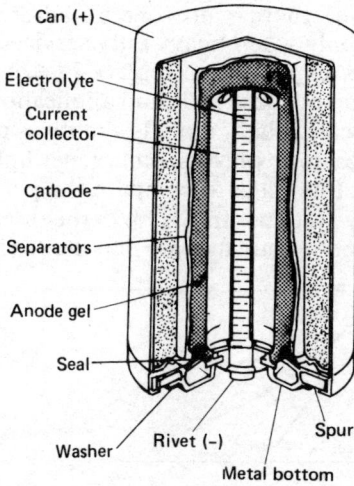

3.21 Cross-section of a D-size alkaline manganese secondary cell. (By courtesy of Union Carbide)

of zinc during recharge. In addition, the cathode mix contains extra binding agents to help preserve its integrity when volume changes occur during charge/discharge cycles.

Performance: primary system

The discharge curve for a D-size alkaline manganese cell with an initial current drain of 500 mA is shown in Fig. 3.22. Compared with even the best

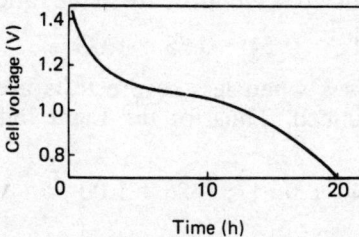

3.22 Discharge curve of a D-size alkaline manganese primary cell with an initial current drain of 500 mA.

Leclanché cell of this size, the alkaline manganese cell has a capacity advantage of at least a factor of four in high current continuous discharge applications. The shelf life of these cells is also good: after a four year storage at 21 °C they retain more than 80% of their initial capacity. At low temperatures, especially at low current drain, their performance is as good as that of the $ZnCl_2$ cell. However because of their higher cost (as with the $ZnCl_2$ cell due to the use of high quality materials and much more sophisticated construction), there is an economic advantage in using them to replace Leclanché cells only when heavy duty service is required: thus for intermittent use, it is not expedient to replace D-size cells below 300 mA current drain. Alkaline manganese cells find applications in devices such as electric shavers, radio-controlled models (motive power and glo-plug ignition), portable tape cassette players, emergency lighting, etc.

It is perhaps useful at this stage to compare the performances of all the Zn/MnO_2-based primary systems. In Fig. 3.23 the discharge curves of four D-size cells on $2.25\,\Omega$ continuous test are reported. Cell (a) is a standard

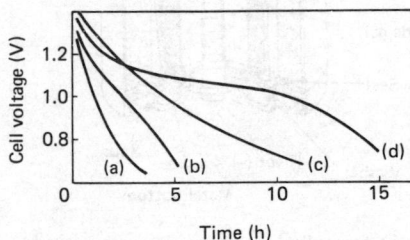

3.23 Comparison of the performances of Zn–MnO_2 primary systems under 2.25 ohm continuous test:
   (a) standard Leclanché cell based on natural ore;
   (b) 'high power' Leclanché cell based on electrolytic MnO_2;
   (c) zinc chloride cell;
   (d) alkaline manganese cell.

Leclanché cell using a natural ore; cell (b) is a 'high power' Leclanché with electrolyte MnO_2; cell (c) is a zinc chloride cell; and cell (d) is an alkaline manganese primary unit. The differences at this current drain are striking: the discharge capacities with a 0.9 V cut-off are in the ratio

$$0.12 : 0.24 : 0.55 : 1.00$$

for the four types. However when less severe tests are considered, the disparities are less pronounced. Thus for the Light Industrial Flashlight (LIF) test, the ratios are

$$0.40 : 0.61 : 0.86 : 1.00$$

Performance: secondary system

Alkaline manganese secondary batteries are generally available as combinations of either D- or G-size cells which have practical capacities of 2.5 and 5.0 Ah respectively. The initial discharge characteristics of the secondary cells are very similar to those of the primary system discussed above. However the cell voltage, and hence the available power and energy, declines as the cycle life of the system runs its course. In Fig. 3.24 the changes in performance of a 15 V battery are shown for a typical charge/discharge schedule. With this type of secondary cell, the cut-off voltages for both discharge and charge are critical. On discharge the cell voltage should

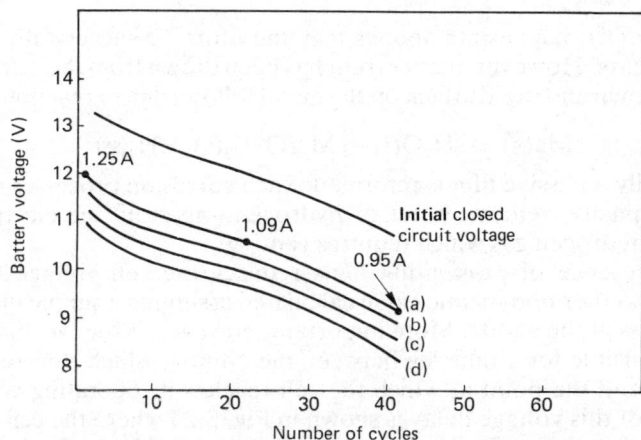

3.24 Performance of a 15 volt alkaline manganese secondary battery as a function of the charge/discharge cycle. Charge: 16 hours voltage limited taper; discharge: 4 hours through 9.6Ω. (a) 1 hour; (b) 2 hours; (c) 3 hours; (d) 4 hours of discharge. (By courtesy of Union Carbide).

not fall below 0.9 V, since otherwise irreversible reduction of the cathode below the nominal $MnO_{1.33}$ may occur. Overcharge has also to be avoided since no provision is made for it, and the use of voltage-limited taper current charging is recommended. Rechargeable alkaline manganese batteries are used in applications where low initial cost is important. The number of cycles obtainable under similar operating conditions is considerably less than for, say, nickel–cadmium batteries, but the cost is much lower. Commercially, this system has had little success.

3.5 Aluminium and magnesium-based Leclanché cells

Replacement of zinc anodes by aluminium or magnesium seems an attractive proposition because of their significantly higher specific capacities (2.98 Ah g^{-1} for Al and 2.20 Ah g^{-1} for Mg compared with 0.82 Ah g^{-1} for Zn). In addition both metals have higher standard potentials than zinc so that a higher cell voltage and energy density can be anticipated. Two problems have delayed the development of practical systems and limited commercial exploitation to a few types of magnesium-based cell: the first is the greatly increased corrosion rate, and the second is the presence of an oxide film which limits anode corrosion, but is responsible for a 'voltage delay' on discharge.

Passivating films

The effect of passivating films on aluminium and magnesium has been the subject of much research. By incorporating chromate/dichromate mixtures and other substances in the electrolyte, a coherent insoluble oxide film is formed which effectively inhibits further corrosion. Sealed cells with aluminium or magnesium anodes may therefore be successfully stored for several years. However once current has been drawn from the cell, the film is broken down and rapid attack on the metal follows due to reactions such as:

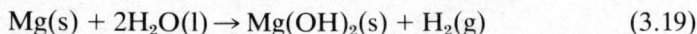

$$Mg(s) + 2H_2O(l) \rightarrow Mg(OH)_2(s) + H_2(g) \qquad (3.19)$$

until finally a passive film is reformed. Such corrosion processes reduce the anode capacity, remove water or hydrogen ions from the electrolyte and produce hydrogen gas which requires venting.

The presence of passivating films reduces the cell voltage below the anticipated thermodynamic value calculated assuming a simple metal/metal ion process at the anode. More important, however, is the fact that the films are responsible for a time lag between the point at which a current drain is initiated and the point at which the cell reaches its operating voltage. An example of this voltage delay is shown in Fig. 3.25 where the cell voltage of an aluminium-based D-size Leclanché cell is plotted as a function of time during the passage of a 0.5 second, 500 mA current pulse. The initial severe polarisation is due to a combination of the low effective exchange current caused by the presence of the film and the ohmic resistance of the film itself.

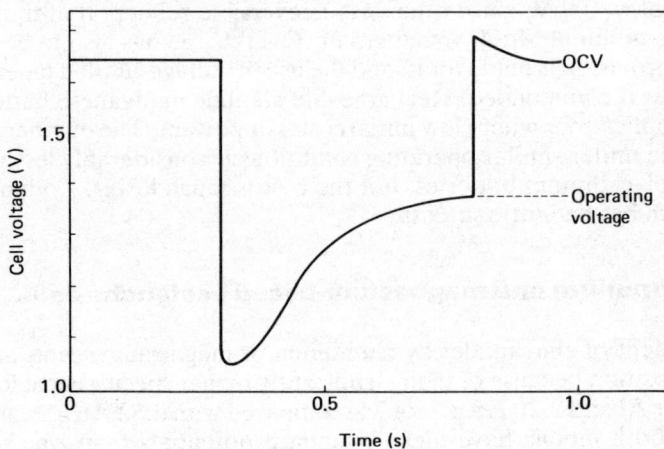

3.25 Voltage delay in an aluminium-based D-size Leclanché cell subjected to a 500 mA pulse for 0.5 seconds.

As the film is disrupted by the passage of current, the cell voltage rises to a steady operating value. The length of the voltage delay is dependent on the anode material, the electrolyte composition and the discharge history, and may extend from a few hundreds of milliseconds to twenty or more seconds. The 'overshoot' in cell voltage when the cell is returned to open circuit corresponds to the period in which a new film is formed: this may be contrasted with the behaviour of lithium in organic solvents (Fig. 5.3). Further consideration is given to voltage delay phenomena in Chapter 5 where lithium anodes are discussed.

Cell constitution

Aluminium-based cells are generally formulated with $AlCl_3$ or $CrCl_3$ solutions as electrolyte. The optimum electrolytes for magnesium-based cells are $MgBr_2$ or $Mg(ClO_4)_2$, buffered with $Mg(OH)_2$. Chromate inhibitors are always added: the exact choice of inhibitor affects the extent of voltage delay phenomena. Resistance to corrosion and shorter voltage delay can also be obtained by using suitable alloys for the anodes. In the case of magnesium, addition of zinc ($\approx 1\%$) reduces the delay and aluminium ($\approx 2\%$) optimises the current efficiency with respect to the corrosion reaction. Similarly for aluminium, a number of alloy compositions and heat treatments have been recommended. The cathode usually resembles closely the MnO_2 system of the standard Leclanché cell. Experimental cells with silver, mercuric and other oxides have been developed but have not been exploited commercially.

Performance

A characteristic feature of cells having aluminium or magnesium anodes is their higher working voltage in comparison with their zinc analogues: thus, the magnesium-based Leclanché cell has an OCV of approximately 1.9 V. The capacity of these cells is however very variable, being dependent on the extent of the corrosion reaction, which is in turn a function of the discharge regimen. For intermittent service, practical capacities as low as 40% are common, whereas high-rate discharge may furnish 70% of the theoretical capacity, and so give a specific capacity of over twice that of a conventional Leclanché cell.

A number of cylindrical and flat magnesium-based cells have been developed on a commercial scale, mainly for military applications where high discharge currents and low unit weight are important. However for most of these applications magnesium batteries have now been replaced by various lithium/organic systems.

3.6 Zinc—mercuric oxide, zinc—silver oxide and related systems

'Miniature' or 'button' cells are cylindrical in form and have a height of less than 5 mm. The range of dimensions encompassed by the term 'miniature' is not clearly defined: it has been suggested* that a good working definition of the term might be 'too small to allow the printing of information such as maker's name, voltage, etc., on the cell case'. The market for these cells has exploded over the past ten years due to the development of electric watches and other miniature electronic devices and current production is over 10^9 units per annum. The various systems all share two favourable features, namely high volumetric capacity, relatively unaffected by current drain, and good discharge characteristics, even under conditions of relatively heavy

* P. Ruetschi (Leclanché S.A., Switzerland) ISE Meeting, Venice 1980.

drain. The earliest aqueous system, based on mercuric oxide and zinc, was introduced in the 1940's. This is the Ruben–Mallory or RM cell, a classic combination of electrochemical and engineering ingenuity which revolutionised the battery industry. The zinc–silver oxide system was introduced commercially by Union Carbide in 1961 shortly after the appearance of electric watches, and a number of other alkaline electrolyte button cells have been developed more recently.

The zinc-mercuric oxide system

This system, commonly known as the 'mercury cell', is based on an amalgamated zinc anode, a concentrated aqueous potassium hydroxide electrolyte – saturated with zincate ion by zinc oxide, and a mercuric oxide/graphite cathode:

$$Zn(s)|ZnO(s)|KOH(aq)|HgO(s),C(s)$$

The anode reaction may be written as

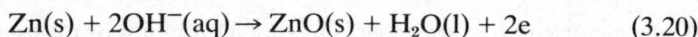

$$Zn(s) + 2OH^-(aq) \rightarrow ZnO(s) + H_2O(l) + 2e \qquad (3.20)$$

and the cathode reaction as

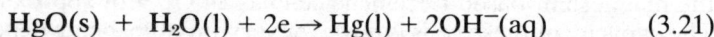

$$HgO(s) + H_2O(l) + 2e \rightarrow Hg(l) + 2OH^-(aq) \qquad (3.21)$$

so that the overall cell reaction is

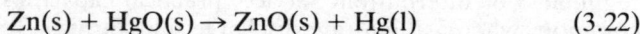

$$Zn(s) + HgO(s) \rightarrow ZnO(s) + Hg(l) \qquad (3.22)$$

Two important points to note are (a) the invariance of the electrolyte solution and (b) the constancy of the chemical potentials of reactants and products, as the discharge proceeds. One consequence of the effective non-involvement of the electrolyte is that only a very small quantity is required in a working cell. Another is a relatively constant internal resistance, leading to a flat discharge curve. The constancy of chemical potentials implies a constant OCV during the course of the discharge, as discussed in Chapter 2. The free energies of formation of $HgO(s)$ and $ZnO(s)$ are given as 58.4 kJ mol^{-1} and 318.2 kJ mol^{-1} respectively in Latimer's 'Oxidation Potentials'. Hence the free energy change for the cell reaction is 259.8 kJ mol^{-1} and the cell e.m.f. is 1.347 V which is in very satisfactory agreement with the OCV of 1.357 V of commercially produced cells. The OCV may be slightly increased by adding MnO_2 to the cathode mix.

The electrolyte is usually an approximately 40% solution of KOH saturated with zinc oxide, to which corrosion inhibitors have been added. The KOH is occasionally replaced by NaOH. Caustic soda solutions have a lower tendency to creep, but have a higher electrical resistance. The electrolyte is immobilised using felted cellulose. The most common anode is a porous compressed cylindrical pellet of amalgamated zinc powder and electrolyte (possibly gelled). An alternative configuration is the 'wound anode' usually found in secondary zinc–mercury oxide cells, which uses a spiral of corrugated zinc foil interleaved with an absorbent paper strip. (The corrugations increase the surface area and provide adequate volume for the deposition of zinc oxide during discharge.) The cathode pellet consists of mercuric oxide together with 5–10% of finely divided graphite, added to increase the electronic conductivity and to minimise the coalescence of mercury formed during discharge. The cathode has always a larger capacity than the anode. The cell is therefore 'zinc limited': i.e. in an exhausted cell there is no zinc left which might corrode and thus lead to hydrogen pressure developing in the cell. A microporous plastic barrier layer is generally placed next to the cathode pellet to prevent internal short circuits caused by the displacement of free mercury or graphite.

The cross-section of a typical mercury button cell is shown in Fig. 3.26. The cathode and anode current collectors are the steel case and steel top respectively. Attention is drawn to the sophisticated engineering design of this cell, which has provision for automatic venting of any pressure caused by hydrogen evolution, with any electrolyte displaced being absorbed in the safety sleeve between the inner and outer case.

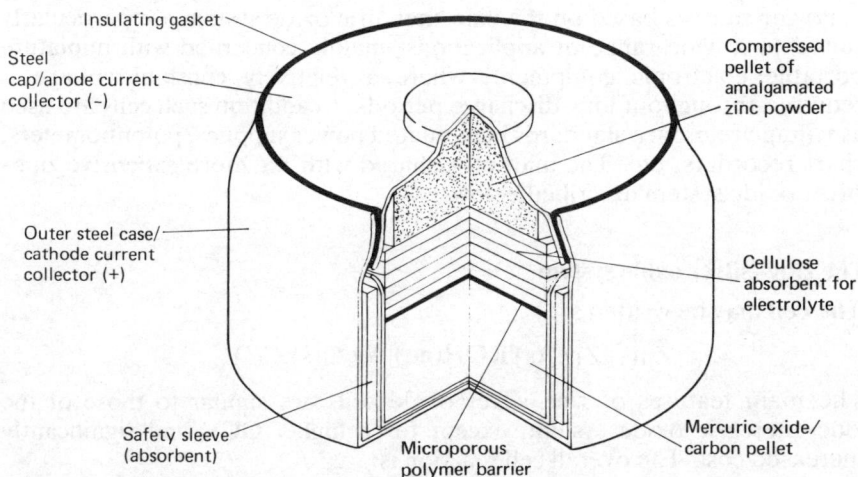

3.26 Cross-section of a typical zinc–mercuric oxide button cell.

Mercury cells have practical specific capacities of up to 400 Ah dm^{-3} and specific energies of 550 Wh dm^{-3}. In addition, they have particularly flat discharge characteristics even under conditions of continuous discharge. These are nearly independent of load over a wide range, as seen in Fig. 3.27. The flat part of the curve is known as the 'equilibrium region' and may extend to 97% of the cell capacity at low current drains (e.g. 1 mA for a 20 mm diameter button cell). Momentary short circuits do not damage the cell and voltage recovery is rapid. Storage behaviour is good, with over 90% of the initial capacity retained after one year. Low temperature performance is not particularly favourable, although it is improved if the wound anode configuration is used, especially if the current demands are low or intermittent.

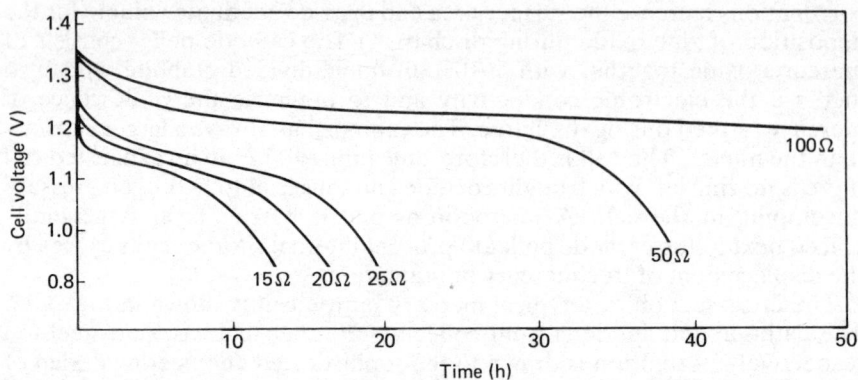

3.27 Discharge characteristics of 1 Ah zinc–mercuric oxide button cell under continuous load at room temperature.

Power sources based on the zinc–mercuric oxide system are particularly suited to a wide range of applications, mainly concerned with miniature portable electronic equipment, where a relatively constant voltage is required throughout long discharge periods. In addition such cells are used as voltage reference standards in regulated power supplies, potentiometers, chart recorders, etc. The market is shared with the more expensive zinc–silver oxide system described below.

The zinc–silver oxide system

The cell may be written as:

$$Zn(s)|ZnO(s)|KOH(aq)|Ag_2O(s),C(s)$$

The main features of zinc–silver oxide cells are similar to those of the zinc–mercuric oxide system, except for a higher OCV and significantly increased cost. The overall cell reaction is:

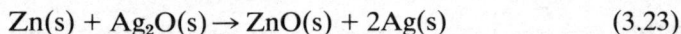

$$Zn(s) + Ag_2O(s) \rightarrow ZnO(s) + 2Ag(s) \tag{3.23}$$

The e.m.f. calculated from Latimer's tables of free energies is 1.593 V which agrees well with the OCV of commercial cells of 1.60 V. A cutaway view of a typical silver oxide button cell is shown in Fig. 3.28: details of the sealing arrangement vary from manufacturer to manufacturer. Discharge curves for a 75 mAh hearing aid battery are shown in Fig. 3.29 for two typical loads. Detailed differences in formulation are made in the production of cells for different uses. For example, electronic watches and other devices with liquid crystal diode (LCD) displays require currents of 3–10 μA and high resistance batteries with NaOH electrolyte are suitable. Watches with light emitting diode (LED) displays or with LCD displays together with additional electrical illumination or alarms, on the other hand, require batteries with low internal resistance which can maintain a stable voltage while supplying current pulses of up to 70 mA for one or two seconds. Cells for such applications use KOH electrolyte. This electrolyte is also used for

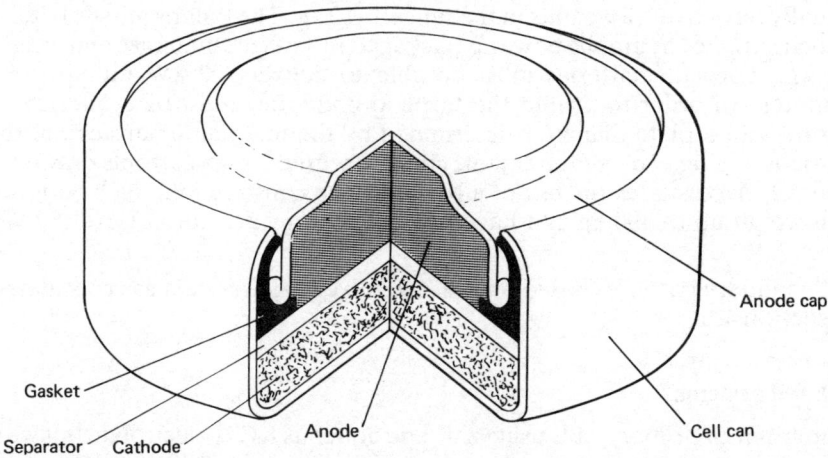

3.28 Cutaway view of a typical zinc–silver oxide button cell. (By courtesy of Union Carbide)

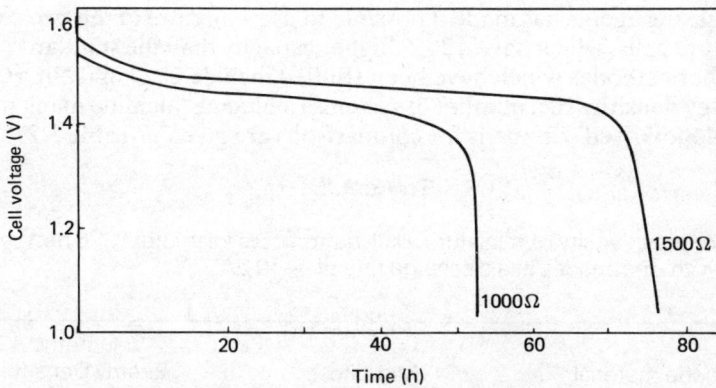

3.29 Discharge characteristics of 75 mAh zinc–silver oxide hearing aid cell under continuous load at room temperature.

hearing aid batteries where continuous current drains of 1–2 mA are typical.

Another application for somewhat larger zinc–silver oxide cells is in the production of high-voltage, high capacity reserve batteries for military equipment: e.g. for missile guidance systems and electrically driven torpedo motors. Such batteries are remotely activated by an electric signal which causes a diaphragm in an electrolyte reservoir to be broken, or electrolyte to be forced into the cells under pressure. The electrolyte solution may be contained in a tubular coil, in cylindrical tanks fitted with pistons or in pressure-over-liquid (POL) tanks. Missile batteries together with their electrolyte delivery systems are carefully assembled and packed in metal containers of sophisticated design and are able to sustain shock levels of up to 2000 G and other extremes of mechanical and environmental stress. They usually have overall weights in the range 1–10 kg. The Polaris missile battery which supplies hydraulic power is packaged in a magnesium case and weighs 14 kg. Torpedo batteries must be able to deliver 300 kW for up to 10 minutes. In order to fit into the torpedo body they must be cylindrical in form, with a plate diameter determined by the internal dimensions of the torpedo casing. This requirement, coupled with the high currents drawn ($\approx$ 500 A), necessitates the use of a bipolar stack construction. Such batteries can weigh up to 400 kg and have practical energy densities of over 80 Wh kg^{-1}.

Secondary batteries based on the zinc–silver oxide system are considered in section 4.6.

Related systems

A number of primary cells using zinc and aqueous KOH electrolyte but with alternative cathode materials have been developed. In 1976 a miniature cell based on a mixture of Ag_2O and AgO was introduced. AgO has a higher energy density and nearly twice the specific capacity of Ag_2O, but it is metastable and tends to decompose by evolving oxygen. However research into stabilising agents has made it possible to use a mixture of the two oxides in miniature cells, which have a 20% higher capacity than the standard Ag_2O cells. Other cathodes which have been studied include CuO and NiO(OH). The energy densities of a number of systems (including 'alkaline manganese' and 'air depolarised' zinc cells for comparison) are given in Table 3.2.

Table 3.2

Practical energy density of miniature alkali button cells (diameter, 11.6 mm; height 4.2 mm) with zinc anodes at a discharge rate of 5–10 μA.

Cathode material	Mid-Life Voltage	Volumetric Energy Density Wh cm^{-3}
MnO_2	1.30	0.23
NiOOH	1.55	0.23
Ag_2O	1.55	0.45
CuO	0.90	0.50
HgO	1.35	0.53
AgO	1.55	0.60
Air	1.25	0.95

The primary objective of miniature battery design is to maximise the energy density in a small container. A compromise must be reached, however, since volumetric energy density decreases as cell volume decreases and the 'dead volume' due to containers, seals, etc., becomes increasingly significant. A plot of energy density as a function of total volume is given in Fig. 3.30 for the zinc–mercuric oxide and zinc–silver oxide systems.

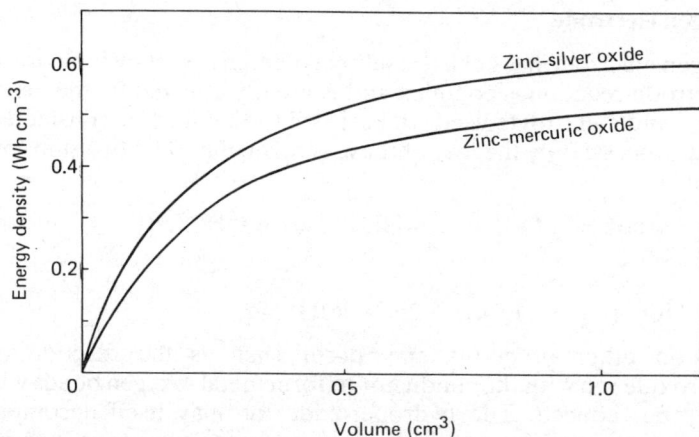

3.30 Energy density as a function of total volume for the zinc–mercuric oxide and the zinc–silver oxide systems. (By kind permission of P. Ruetschi, Leclanché S.A., Yverdon, Switzerland)

3.7 Metal-air batteries

A number of cells have been developed which make use of the oxygen of the air as the cathodic reactant. These so-called 'air-depolarised cells' are an example of a **hybrid cell** discussed more fully in Chapter 8. Many recent advances in metal–air batteries can be attributed to the research carried out in the 1960's on high current density air electrodes for ambient hydrogen/oxygen fuel cells using aqueous electrolytes.

The most obvious advantages of the oxygen cathode is that it has low weight and infinite capacity. Consequently, prototype D-size cells based on the zinc–air system have been shown to have twice the overall practical capacity of zinc–mercuric oxide cells (and ten times that of a standard Leclanché cell) when subjected to a continuous current drain of 250 mA. In the larger industrial cells energy densities of up to 200 Wh kg^{-1} and specific capacities of 150 Ah dm^{-3} may be obtained. On the other hand, a catalytic surface must be provided for efficient charge transfer at the oxygen cathode, and by its nature, the electrode is susceptible to concentration polarisation.

The history of metal–air batteries goes back over one hundred years to the work of Maiche who modified a Leclanché cell by replacing the conventional MnO$_2$ cathode with a mixture of platinum and carbon powder. Successful commercial primary cells have ranged from the well known 500 Ah zinc–air wet cell developed for railway signalling in the 1930's to the more modern hearing aid batteries and high capacity/high current industrial primary systems. Secondary hybrid cells using air electrodes will be considered in Chapter 8.

The oxygen electrode

The oxygen electrode has been the subject of intensive study for many years. The electrode reaction is complex and is greatly affected by the electronic conductor and electrolyte used. In basic solution it may be considered as a two stage process: only the first of these is reversible. The two steps may be written as:

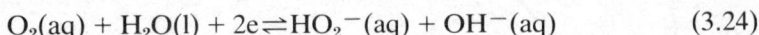

$$O_2(aq) + H_2O(l) + 2e \rightleftharpoons HO_2^-(aq) + OH^-(aq) \qquad (3.24)$$

and

$$HO_2^-(aq) + H_2O(l) + 2e \rightarrow 3OH^-(aq) \qquad (3.25)$$

In addition other processes may occur such as the reaction of the hydroperoxide ion with the conductor to form metal–oxygen bonds which in turn may be reduced. The hydroperoxide ion may itself decompose to reform oxygen, etc. The potential of an oxygen electrode is invariably a mixed potential with a value of about 1.0 V on the standard hydrogen scale, at zero current drain.

Since the electrode reaction can only occur in the region where solid, liquid and gaseous phases come together, the construction of oxygen electrodes for practical cells is designed to maximise the interfaces between them. This may be achieved, for instance, by using porous nickel or carbon treated with metal or metal oxide catalysts (Kordesch electrodes). The pores are made accessible to both electrolyte solution and air (Fig. 3.31) and to prevent flooding and eventual leakage of the former, the surfaces of the electrode which are exposed to the air are impregnated with a water-repellent coating using paraffin wax or a synthetic polymer. It should be noted that oxygen electrodes of this type would be able to pass much larger currents if they were supplied with pure oxygen rather than air.

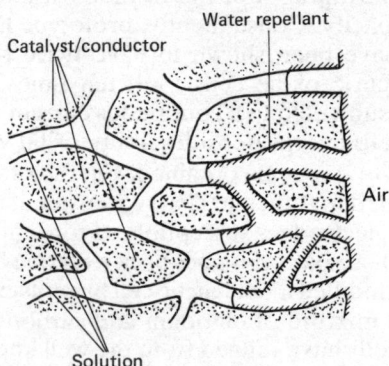

3.31   Schematic view of the interphase at a porous matrix air electrode.

The electrolyte for zinc-based cells is always caustic alkali. Calcium hydroxide is sometimes added to remove zinc ions as insoluble $CaZn_2O_3.5H_2O$. A caustic alkali electrolyte is effectively buffered against

OH$^-$ ion production by the oxygen cathode, so that OH$^-$ concentration polarisation is not serious. On the other hand, such an electrolyte can readily become contaminated with carbonate by reaction with the carbon dioxide of the air. NaOH is substituted by KOH in zinc–air cells intended for low temperature use. Oxygen electrodes used with the near neutral electrolytes suitable for aluminium and magnesium anodes are more subject to polarisation and are limited to lower current drain applications. Further details concerning oxygen electrode polarisation are given in Section 8.2.

Anodes

Four metals have been studied extensively for use in this type of system, namely zinc, aluminium, magnesium and lithium. However the last three metals suffer from severe corrosion problems during storage and magnesium–air and aluminium–air cells are generally operated either as 'reserve' systems in which the electrolyte solution is added to the cell only when it is decided to commence the discharge, or as 'mechanically rechargeable' batteries which have replacement anode units available. The lower energy and power density of zinc is compensated for by the ease with which serious corrosion may be inhibited, so that zinc is by far the most commercially important anode in primary metal–air cells.

In early wet cells, lightly amalgamated solid zinc plates were used. Fig. 3.32 shows a cross-section of such a cell: the tapered shape of the zinc plates allowed for the higher current densities near the top of the air

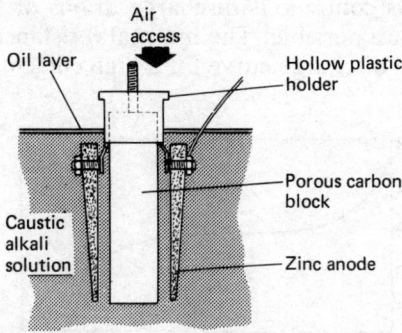

3.32 Cross-section of the electrode assembly of an early 500 Ah mechanically rechargeable zinc-air wet cell, used for railway signalling applications.

electrode. In basic solution, unsaturated with zincate ions, the anode reaction may be written as:

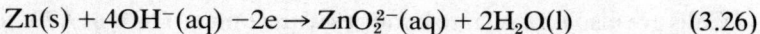

$$Zn(s) + 4OH^-(aq) - 2e \rightarrow ZnO_2^{2-}(aq) + 2H_2O(l) \qquad (3.26)$$

When the solution becomes saturated with zincate, zinc oxide is formed:

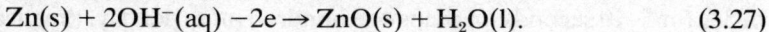

$$Zn(s) + 2OH^-(aq) - 2e \rightarrow ZnO(s) + H_2O(l). \qquad (3.27)$$

A number of techniques have been used to prevent degradation of battery performance caused by zinc oxide passivation. In the early wet cells, despite the penalty of reduced energy density, sufficient electrolyte was added to allow most of the zinc to dissolve. In more modern construction, anodes are made by compacting powdered zinc onto brass current collectors or by electrolytic reduction of pasted sheets, to form a porous mass with a high area/volume ratio. In this configuration the oxide does not significantly block further oxidation of the zinc. In addition, most of the electrolyte required may be incorporated within the pores. A carefully positioned current collecting grid is a necessity for such a cell since, by its nature, it is always anode limited.

The pros and cons of aluminium and magnesium anodes were discussed in section 3.5. The corrosion problem is even more serious in metal–air cells since the electrolyte may be saturated with oxygen.

Typical cells

Zinc-based industrial primary cells range in size from 90 Ah cylindrical cells used mainly in telecommunications, hazard warning lights, etc., to 2000 Ah cells designed for inshore navigation beacons, standby power and railway track and signalling circuits.

The cell may be written as

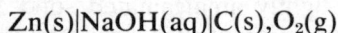

$$Zn(s)|NaOH(aq)|C(s),O_2(g)$$

and the nominal OCV is 1.4 V.

With the larger cells continuous discharge drains of 1A and intermittent discharges of 2.5A are possible. The internal resistance of such cells is low ($\approx 0.1\ \Omega$). A typical discharge curve for a high current/high capacity cell is shown in Fig. 3.33.

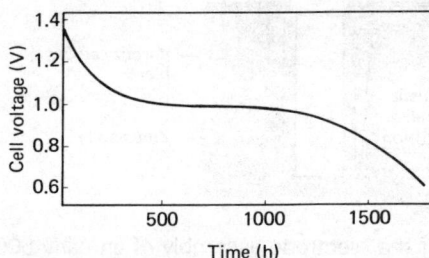

3.33 Continuous discharge curve through 2 ohm load of a 750 Ah zinc–air cell. Such cells have a 5A intermittent pulse capability and good capacity retention. (By courtesy of SAFT (U.K.) Ltd.)

Cells are also formed into batteries with nominal OCV of 7.2 V for electric fence activators. Alternatively, banks of eleven individual cells are used to operate 10 V railway signal motors: such applications require a current drain of 3A for 5–10 seconds, perhaps one hundred times per day. Practical energy densities of up to 310 Wh kg^{-1} may be obtained.

A number of commercial zinc–air cells are available in reserve form:

usually they require simply the addition of clean fresh or sea water to activate them.

In Fig. 3.34 a miniature 'air-depolarised' zinc cell is shown. Such cells are used for hearing aids and similar devices: they have a similar configuration to the more conventional mercury or silver button cell, but a much higher capacity for the same volume – e.g. 560 Ah dm^{-3} (190 Ah kg^{-1}).

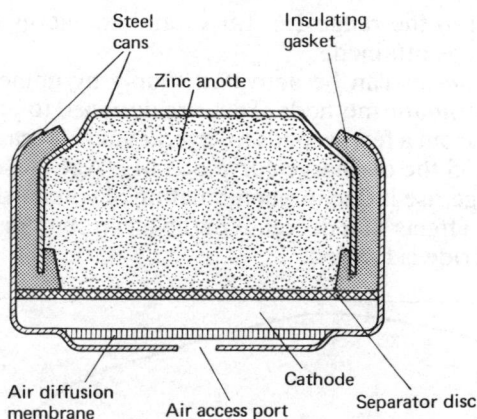

3.34 Cross-section of zinc–air button cell. (By courtesy of Gould)

Zinc–air cells with electrolytes based on NH_4Cl and a similar OCV are also manufactured:

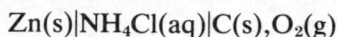

$$Zn(s)|NH_4Cl(aq)|C(s),O_2(g)$$

The cell reaction can be written as

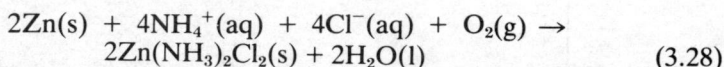

$$2Zn(s) + 4NH_4^+(aq) + 4Cl^-(aq) + O_2(g) \rightarrow$$
$$2Zn(NH_3)_2Cl_2(s) + 2H_2O(l) \tag{3.28}$$

Magnesium–air batteries have been developed as 'mechanically recharge-able' reserve batteries in modular form. In the General Electric multicell system the air electrodes are integral with the battery casing, but the magnesium alloy anodes slot into guides and may be readily replaced. The electrolyte is sea water or a 7% NaCl solution. The so-called 'magnesium-inert cathode' cell is in practice a low efficiency magnesium–air cell. In this system the cathode is nickel-plated iron and the anode/cathode assembly is immersed in sea water to provide low currents in long term low maintenance applications. No commercial aluminium–air battery has yet been marketed.

3.8 Magnesium reserve batteries

In previous sections of this chapter magnesium anodes have been considered as replacements for zinc in Leclanché and air-depolarised cells. In sea water activated reserve batteries magnesium anodes are coupled with either silver chloride, lead chloride or, occasionally cuprous chloride. In the case of silver chloride the cell may be represented as:

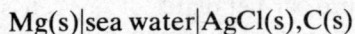

$$Mg(s)|sea water|AgCl(s),C(s)$$

and the cell reaction as:

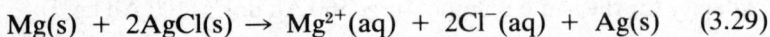

$$Mg(s) + 2AgCl(s) \rightarrow Mg^{2+}(aq) + 2Cl^-(aq) + Ag(s) \quad (3.29)$$

OCV values fall in the range 1.6–1.9 V, and an energy density as high as 100 Wh kg^{-1} can be attained.

Sea water batteries can be activated simply by immersion or by more sophisticated automatic methods. They are designed to provide currents for periods ranging from a few seconds to several hours (occasionally one or two days). In Fig. 3.35 the discharge curve of a Mg–AgCl reserve cell is shown. The initial voltage rise is due to a combination of voltage delay at the anode, followed by the effects of a decrease in the electrolyte resistance as soluble magnesium chloride is formed.

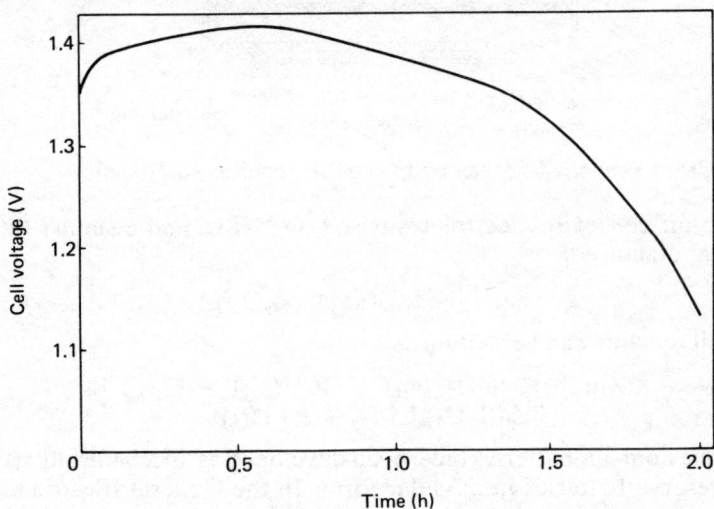

3.35 Discharge curve for sea water activated magnesium–silver chloride reserve cell.

Small cells are used to illuminate air-sea rescue beacons on life-jackets and life-rafts. Larger battery units with working voltages ranging up to several hundred volts are used for a variety of maritime applications, including provision of motive power for torpedo motors.

4 Secondary aqueous electrolyte cells

4.1 Introduction

The manufacture of secondary batteries based on aqueous electrolytes forms a major part of the world electrochemical industry. By far the largest total capacity produced is that of lead–acid batteries whose dominance is due to a combination of low cost, versatility and the excellent reversibility of the electrochemical system. Lead–acid cells have extensive use both as portable power sources for vehicle service and traction, and in stationary applications ranging from small emergency supplies to load levelling systems. The nickel–cadmium cell shares the market for portable power, but accounts for only a small percentage of total sales. It is more expensive, but has a number of advantages such as robustness, low maintenance requirements, and good high rate and low temperature characteristics. Other systems, although having only minor commercial importance, may be preferred for special applications, for instance where the high relative atomic mass of lead and its consequent low energy density are a particular disadvantage – such as in EV traction.

4.2 The lead–acid cell

Introduction

The history of the lead–acid cell, more consistently named the 'lead–lead oxide cell', commenced in 1859 with the construction by the French physicist, Gaston Planté, of the first practical rechargeable cell, consisting of two coiled lead strips, separated by a linen cloth (Fig. 4.1). This system forms the basis of the most widely used secondary battery of the present time. Today over a third of the world output of lead is used to manufacture over one hundred million lead–acid batteries every year. These range in size from 5 Wh cells to 100 Wh starting, lighting and ignition (SLI) systems, and to projected 10 MWh load levelling modules. The great success of this system is due to a number of favourable factors such as the relatively low cost and availability of the raw materials (lead and sulphur), ease of manufacture, long cycle life and favourable electrochemical characteristics. The performance of the lead–acid cell has been improved through more than a century of continuous research and development. This has been directed not only towards improved engineering design and manufacturing techniques,

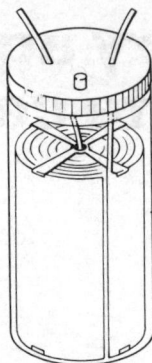

4.1   Original Planté spiral wound lead–acid cell.

but also to achieving a better understanding of the mechanisms of the cell process and of the factors which play an important role in degrading the system during its cycle life. The result has been a significant increase in the specific capacity, energy density and cycle efficiency and a reduction in the need for cell maintenance – e.g. low rate charging to compensate for self-discharge.

The lead–acid cell can be represented schematically as having a negative electrode of porous lead (lead sponge) and a positive electrode of lead dioxide, PbO_2, both immersed in an aqueous solution of sulphuric acid:

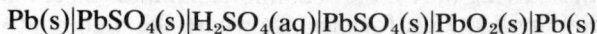

$$Pb(s)|PbSO_4(s)|H_2SO_4(aq)|PbSO_4(s)|PbO_2(s)|Pb(s)$$

The overall electrochemical processes can be represented by the equation

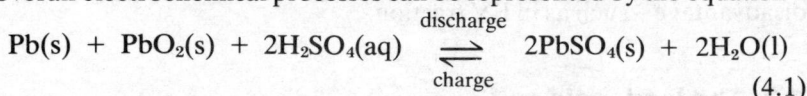

$$Pb(s) + PbO_2(s) + 2H_2SO_4(aq) \overset{\text{discharge}}{\underset{\text{charge}}{\rightleftharpoons}} 2PbSO_4(s) + 2H_2O(l)$$

$$(4.1)$$

As the cell is discharged, sulphuric acid is consumed and water is formed. Consequently the electrolyte composition and density vary from about 40% by weight of H_2SO_4 (1.30 kg dm^{-3}) at full charge, with an associated OCV of 2.15 V at 25 °C, to about 16% by weight of H_2SO_4 (1.10 kg dm^{-3}) when fully discharged, with an OCV of 1.98 V. The change in electrolyte specific gravity provides a convenient method of determining the state of charge of a cell. The open circuit voltage depends on the sulphuric acid (and water) activity and temperature and may be predicted with accuracy from thermo-dynamic free energy values. In Fig. 4.2 the OCV is given as a function of the percentage of discharge capacity. The discharge process also results in the formation of insoluble lead sulphate on both electrodes. This material is a very poor electrical conductor and its deposition in a dense, fine-grained form can shield and passivate both electrodes, so that the practical capacity of a cell can become severely restricted – to as little as 5–10% of the theroretical capacity for large current densities. A number of techniques for reducing this effect will be described below. As discharge proceeds, the internal resistance of the cell rises, due to $PbSO_4$ formation and the decrease in electrolyte conductivity as H_2SO_4 is consumed. During charge, $PbSO_4$ is

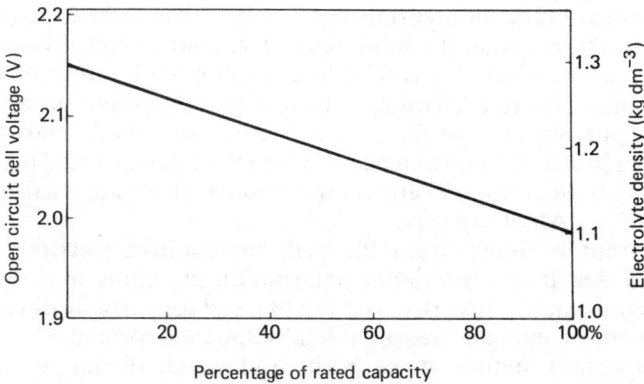

4.2 Approximate open circuit voltage and electrolyte density as a function of percentage service capacity for the lead–acid cell.

reconverted to lead at the negative* and to PbO_2 at the positive*. The energy efficiency of the charge/discharge cycle may be high, but depends on charge rates and cell design. In particular, it is affected by a side reaction in which oxygen is evolved at the positive during charging. In some applications it is standard practice to supply at least 10% overcharge to the battery in order to obtain the full discharge capacity of the positive. This overcharge leads to hydrogen formation at the negative and further oxygen evolution at the positive plate, and thus causes water loss from the electrolyte.

Positive electrodes

The electrochemical reactions at the positive electrode are usually expressed

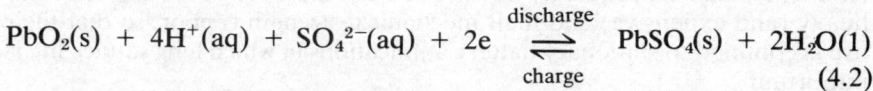

$$PbO_2(s) + 4H^+(aq) + SO_4^{2-}(aq) + 2e \underset{\text{charge}}{\overset{\text{discharge}}{\rightleftharpoons}} PbSO_4(s) + 2H_2O(l) \tag{4.2}$$

In practice the bisulphate ion, HSO_4^-, is a rather weak acid ($pK_a = 1.99$ at $25\,°C$), so that for the sulphuric acid concentrations used in practical cells the reactions

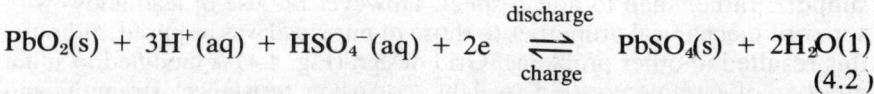

$$PbO_2(s) + 3H^+(aq) + HSO_4^-(aq) + 2e \underset{\text{charge}}{\overset{\text{discharge}}{\rightleftharpoons}} PbSO_4(s) + 2H_2O(l) \tag{4.2′}$$

may be considered a more accurate description of the electrode process. Both lead dioxide and lead sulphate are slightly soluble, and it is probable that soluble lead species are involved in the reaction mechanism. In the absence of mass transport limitations, it is considered that the process.

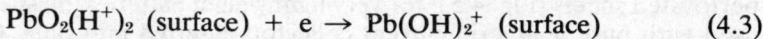

$$PbO_2(H^+)_2 \text{ (surface)} + e \rightarrow Pb(OH)_2^+ \text{ (surface)} \tag{4.3}$$

is rate determining for the discharge reaction. The lead dioxide is found to vary slightly in stoichiometry at different stages in the cycle. Immediately after charge it may have a composition as high as $PbO_{2.05}$.

---

¹ * 'Negative' and 'positive' are the almost universal terms used to identify the **negative electrode** and **positive electrode** respectively of a secondary cell.

In order to obtain optimum current densities, it is necessary to use a highly porous structure so that the solid/electrolyte contact area is large. The fully charged positive electrode is therefore composed of a mass of small PbO_2 crystals connected to each other to form a continuous porous network. The electrode porosity is important for another reason, since it makes allowance for the increase in volume that occurs when PbO_2 is converted to $PbSO_4$. In a typical SLI battery the real area of the positive electrode is calculated to be 50–150 m² per Ah of capacity.

As current is drawn from the cell, the positive electrode voltage is depressed due to concentration polarisation as sulphuric acid within the pores is consumed. This effect is more marked for partially discharged cells since the pore volume decreases as lead sulphate is formed.

An important feature of the positive electrode discharge concerns the nature of the $PbSO_4$ deposit since the formation of dense, coherent layers can lead to rapid electrode passivation. Lead dioxide exists in two crystalline forms, rhombic (α–) and tetragonal (β–), both of which are present in freshly formed electrode structures. Since $PbSO_4$ and α–PbO_2 are isomorphic, crystals of lead dioxide of this modification tend to become rapidly covered and isolated by lead sulphate and their utilisation is less than that of the tetragonal β–form. As the latter is the thermodynamically more stable of the two, some transformation of α– into β–PbO_2 may occur during the life of a battery, with consequent improvement in its performance.

Positive electrodes are manufactured in three forms, as (i) Planté plates, (ii) pasted plates and (iii) tubular plates. In Planté plates, the positive active material is formed by electrochemical oxidation of the surface of a cast sheet of pure lead to form a thin layer of PbO_2. The plate generally has a grooved surface to increase its surface above its geometric area by a factor of 3–10, (Fig. 4.3). Such plates have a very long life since they have a large excess of lead which can subsequently be oxidised to PbO_2. However they are very heavy (and expensive) and their mechanical strength is poor, so that their use is confined to stationary battery applications in which long service life is important.

In 1880, Fauré proposed coating the lead sheet with a 'paste' of lead dioxide and sulphuric acid in order to increase the capacity of the system. It was soon found that the paste would be more readily applied to an open grid support, rather than to a lead sheet. However the use of lead alloys with superior mechanical properties to those of pure lead was required, although this resulted in other problems. Grid design (Fig. 4.4) is modified to suit a number of parameters (e.g. weight, corrosion resistance, strength, and current distribution) which are important in different ways for different battery applications. Since the melting point of lead is low (327 °C), most grids are formed by melting and casting; some lighter varieties are now manufactured using a stretching process which produces 'expanded metal' perforated sheets (Fig. 4.5); for very light weight batteries, plastic grids are used with pure lead strip current collectors inserted into slots. Grids are designed to ensure a low internal resistance for the cell and to minimise shedding of active material on cycling. Shedding causes loss of capacity, and dislodged material can accumulate on the battery case floor where it can give

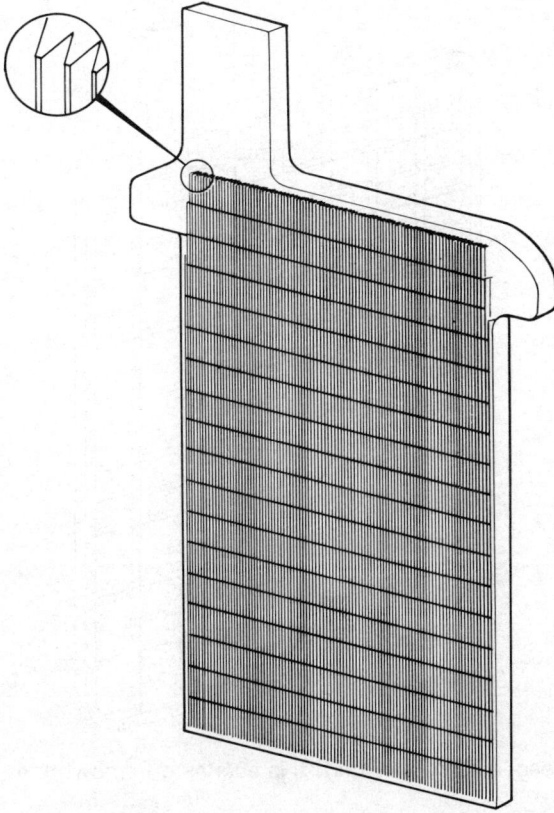

4.3 Planté plate. Detail shows the grooves which increase the effective area of the plate.

rise to short circuits between positive and negative plates. Additions of small quantities of tin are made to the lead to improve its coating properties, while antimony, calcium or selenium are added to form alloys with better stress resistance. Lead–antimony was the first alloy used and still remains the most popular. Antimony (1.5–8%) greatly improves the mechanical properties of grids and connector bars, but also increases their electrical resistance, accelerates the self-discharge of the cell, and reduces cycle efficiency. Further, during recharge, poisonous SbH_3 gas can be formed. Research is continuing into the development of better alloys. Lead–calcium grids are superior in many ways. The use of selenium has allowed the manufacture of low-antimony grids of adequate mechanical strength. The main component of the paste used to fill the grid is known as lead dust, and consists of a carefully milled mixture of metallic lead and lead oxides. Water and sulphuric acid are added in a predetermined sequence, together with minor components and strengthening fibres, and the resulting slurry is loaded onto the grids and 'cured' or dried to produce a crack-free plate, with good adherence to the grid. The active material is later converted to the fully

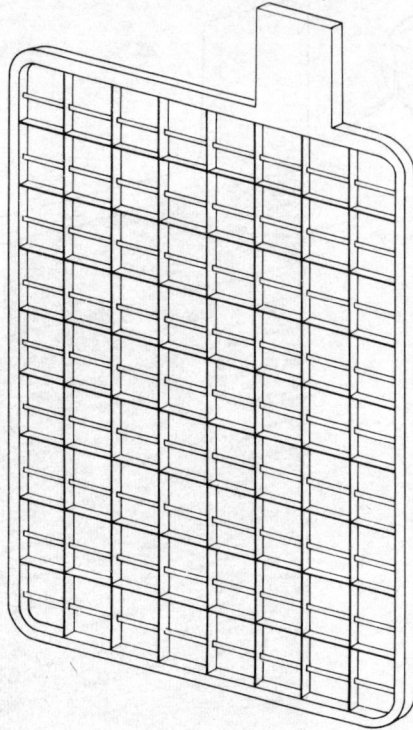

4.4 Typical lead–acid battery grid: this acts as a framework to hold the active material in place.

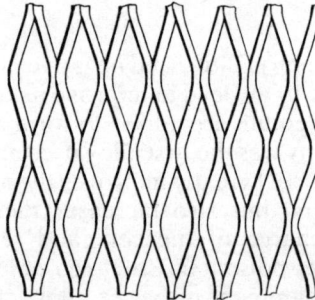

4.5 Expanded metal grid.

charged condition by the process of **forming** as described below. Pasted plates have a relatively high capacity and power density, but are not mechanically strong. They are used extensively in SLI and similar batteries.

Tubular plates (also known as 'armoured' or 'clad' plates) consist of a row of tubes containing axial lead rods surrounded by active material (Fig. 4.6).

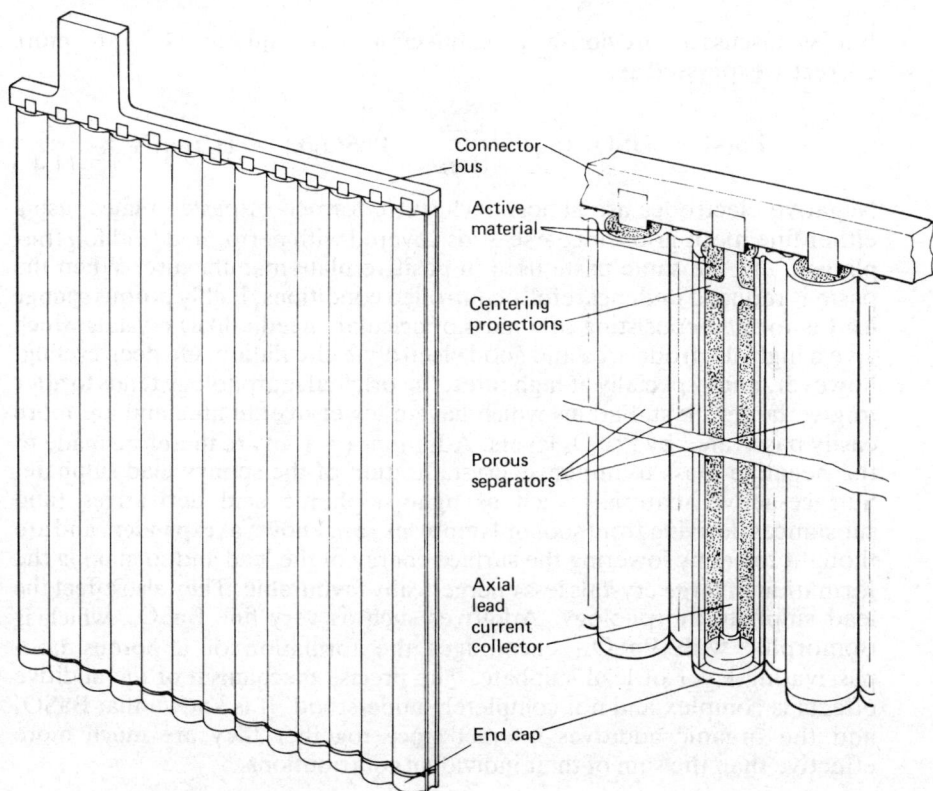

4.6   (a) Tubular plates for lead–acid cells.
       (b) Cross section showing central lead current collector, active material and porous separators.

The tubes are formed of fabrics such as terylene or glass fibre or of per-forated synthetic insulators which are permeable to the electrolyte. They generally have a circular cross-section, but square, rectangular and oval tubes are also manufactured. Tubular plates are sufficiently strong to withstand continuous vibration and are resistant to shedding. They are also able to sustain many deep discharges without loss of integrity and are therefore suitable for applications such as EV traction.

Negative electrodes

The reactions of the negative electrode are generally given as:

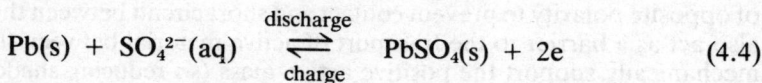

$$Pb(s) + SO_4^{2-}(aq) \underset{\text{charge}}{\overset{\text{discharge}}{\rightleftharpoons}} PbSO_4(s) + 2e \qquad (4.4)$$

but as discussed previously in connection with equation 4.2, are more correctly expressed as:

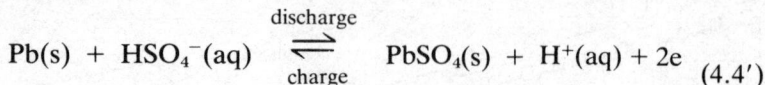

$$Pb(s) + HSO_4^-(aq) \underset{charge}{\overset{discharge}{\rightleftharpoons}} PbSO_4(s) + H^+(aq) + 2e \quad (4.4')$$

Negative electrodes are almost exclusively formed of pasted plates, using either fine mesh grids or coarse grids covered with perforated lead foil (**box plates**), and the same paste used in positive plate manufacture. When the paste is reduced under carefully controlled conditions, highly porous **sponge lead** is formed consisting of a mass of acicular (needle-like) crystals which give a high electrode area and good electrolyte circulation. On deep cycling, however, and especially at high rates, the original morphology tends to alter to give larger crystal grains which have a lower overall area and are more easily passivated by $PbSO_4$ layers. Additions ($\approx 1\%$) are therefore made to the negative mass to minimise crystallisation of the spongy lead sulphate. Surface-active materials such as ligno-sulphonic acid derivatives (and substances deriving from soot or lampblack) are known as **expanders** and are thought to act by lowering the surface energy of the lead and so making the formation of large crystals less energetically favourable. They also affect the lead sulphate morphology. Additives such as very fine $BaSO_4$, which is isomorphic with $PbSO_4$, encourages the formation of a porous non-passivating layer of lead sulphate. The precise mechanism of the additive effects is complex and not completely understood. It is known that $BaSO_4$ and the organic additives interact since together they are much more effective than the sum of their individual contributions.

Electrode forming

Forming is defined as the procedure undertaken, usually before final assembly in the battery case, to convert the active material in the positive and negative plates into their fully charged condition. In effect, the forming process is equivalent to the first charge, but the electrolysis operation is carried out using carefully controlled conditions of temperature and electrolyte composition, and at low current density (typically <100 A m^{-2}), in order to achieve complete conversion throughout the electrode masses while retaining adherence and porosity.

SLI batteries are nowadays often supplied in a dry charged state and are activated simply by filling with electrolyte. Plates for such batteries have extra additives, such as antioxidants in the negative active mass, and forming is followed by one of a number of controlled drying processes.

Separators

Separators are porous insulating sheets which are placed between electrodes of opposite polarity to prevent contact and short circuit between them. They also act as a barrier to the transport of active material between the plates, mechanically support the positive active mass (so reducing shedding) and prevent dendrite formation. Effective separators must possess high porosity so that they will have low electrical resistance, low pore diameter in order to achieve good separation and finally, resistance to oxidation and stability in highly acid conditions. In early batteries, separators were made from fir or cedar wood by removing the resinous substances in order to make them acid resistant. Better results were obtained with microporous rubber separators: natural latex becomes highly porous after vulcanisation due to water evaporation. However present day batteries almost all use separators made from synthetic polymers. The porosity characteristics of these substances can be controlled to within close tolerances during the manufacturing process. The most popular materials include sintered polyvinylchloride, and extruded polyethylene. Special papers, impregnated with phenolic resins are also in current use. Research is continuing to develop separators which are able to retain very large quantities of electrolyte for use in sealed batteries which can be operated in any orientation. Very efficient thin separators are required for high energy density batteries.

Hard rubber or glass fibre is used to fabricate **retainers** which are perforated sheets in contact with the positive plate which protect the separators from its strong oxidising environment.

Final assembly

Plates which have been tank-formed are first separated and cut to size. Lugs are milled free of oxide in preparation for welding to lead connector straps. The plates are assembled into parallel groups or stacks (usually with one extra negative plate), which are then interleaved, with separators, retainers and spacers inserted (Fig. 4.7). In a battery containing more than one cell group (Fig. 4.8) series connections must also be made. In large stationary batteries and older SLI models, these are made using external straps. In modern SLI batteries such connections are made within the battery case using **through-partition ties,** thus saving lead and reducing the internal resistance and weight of the battery.

Cases are now almost exclusively fabricated by injection moulding using synthetic polymers which have replaced the bitumen and hard rubber widely used in the past. Polypropylene has excellent mechanical and chemical properties and allows light, thin-walled monobloc containers to be constructed. Glass containers are still sometimes used for emergency batteries because they permit a rapid check of battery condition, but they are tending to be replaced by glass–polyester materials with antiacid internal coating. Almost all lead–acid battery cases are rectangular (prismatic) in shape. The interior may be manufactured with projections to locate the plates and support them above the case floor, in order to leave **mud** or **sludge** space where shed materials can accumulate (Fig. 4.9). In other batteries the plates are suspended from the case lid or from a ledge in the case wall. The lid is welded or sealed to the case and is provided with apertures for terminal pillars, venting valves, and simple screw caps or complex automatic systems for adding distilled water to the electrolyte.

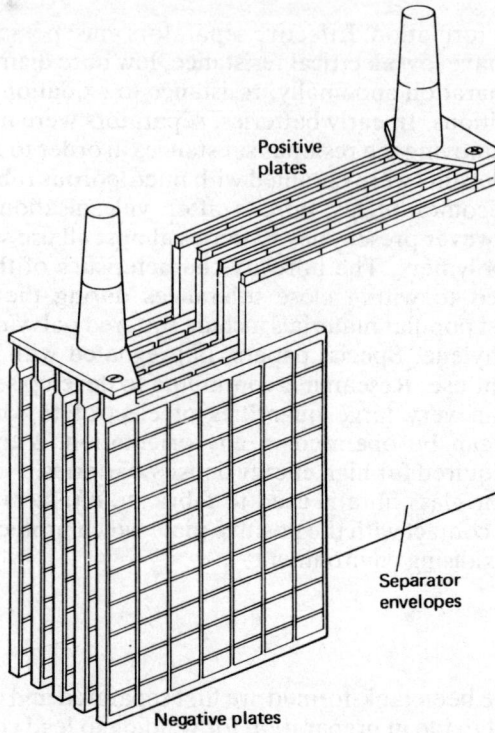

4.7 Interleaving of positive and negative electrode groups (elements) to form a lead–acid cell.

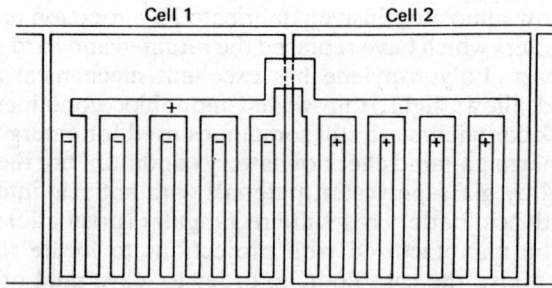

4.8 Schematic diagram of lead–acid battery showing through-partition connection.

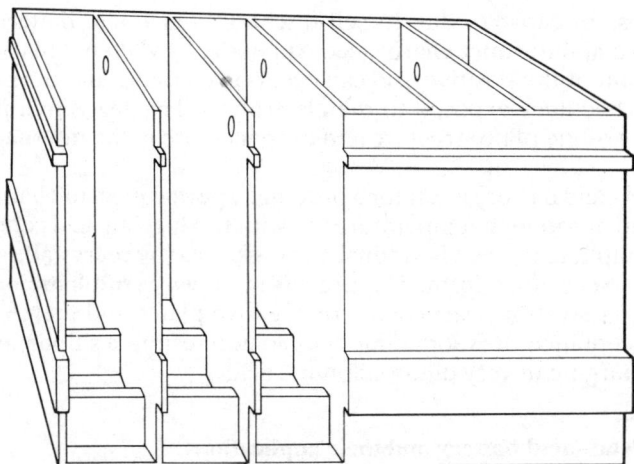

4.9 Monobloc injection moulded case.

Performance

Lead–acid batteries are designed for a wide range of applications, each of which has its own requirements and typical discharge pattern. For example, SLI batteries require short, very high rate discharges (at least 5 C) but are rarely discharged to any great depth. In contrast, EV traction batteries and some stationary industrial installations must be able to sustain deep discharges at effectively constant current (0.1–0.2 C, say). It is therefore not possible to discuss a generalised cycle performance for this system, and instead some details on the performance of certain important types are discussed below.

The practical capacity of all such batteries is also dependent on the temperature of operation and is found to drop very rapidly at temperatures below 0°C.

As the system is thermodynamically unstable with respect to hydrogen and oxygen evolution, lead–acid cells are subject to self-discharge:

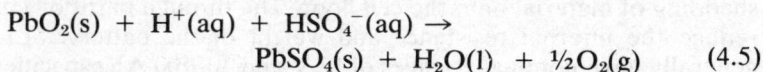

$$PbO_2(s) + H^+(aq) + HSO_4^-(aq) \rightarrow$$
$$PbSO_4(s) + H_2O(l) + \tfrac{1}{2}O_2(g) \qquad (4.5)$$

and

$$Pb(s) + H^+(aq) + HSO_4^-(aq) \rightarrow PbSO_4(s) + H_2(g) \qquad (4.6)$$

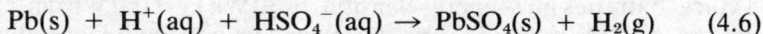

The rates of these processes are dependent on temperature, electrolyte composition, and most importantly, impurity content. If antimony is leached out of the positive grid it may be deposited on the negative plate where it catalyses reaction 4.6, because of its relatively low hydrogen overvoltage. The use of low antimony grids and antimony trapping separators reduces the amount of self-discharge. Reaction of the positive plate material with other solution impurities such as Fe^{2+} ions which can be re-reduced at the negative leads to very rapid self-discharge. To compensate for the loss in capacity due to self-discharge reactions, batteries may be placed on a **maintenance charge** (see Appendix 1) when not in use.

Corrosion of the positive grid can occur on charging and over-charging if the metal becomes exposed to the electrolyte. This leads to a progressive weakening of the plate structure and to an increase in the internal resistance of the cell.

If a lead–acid battery is left for a prolonged period in an uncharged state or is operated at too high temperatures or with too high an acid concentration, the lead sulphate deposit is gradually transformed by recrystallisation into a dense, coarse-grained form. This process is known as **sulphatation** and leads to severe passivation, particularly of negative plates and therefore inhibits charge acceptance. It is sometimes possible to restore a sulphatated battery by slow charging in very dilute sulphuric acid.

Types of lead–acid battery and their applications

Three main types of battery will be briefly described: SLI batteries, industrial batteries (traction and stationary) and small sealed portable batteries. The order follows their present relative commercial importance.

SLI batteries

As is well known, these batteries are used for cranking automobile internal combustion engines, and for supporting devices which require electrical energy when the engine is not running. The major part of the growth of the lead–acid battery industry in recent years is related to the world increase in the number of cars and lorries. About 80% of all lead acid battery production goes to supply this market.

SLI batteries must be capable of supplying short but intense discharge currents at rates of over 5 C. They are therefore generally constructed of thin pasted plates, with thin composite separator/retainer layers and short connector buses to minimise the internal resistance. A typical battery is shown in Fig. 4.10. In this unit, the positive plates are inserted into pocket shaped separators to increase their resistance to shock and prevent the shedding of material onto the cell floor. The through partition connections reduce the internal resistance and weight of the battery. SLI batteries generally have nominal voltages of 12 V and 30–100 Ah capacities for cars, and 24 V and up to 600 Ah for lorries, construction and military vehicles. Typical batteries have energy densities of 30 Wh kg^{-1} (60 Wh dm^{-3}) but units with up to 40 Wh kg^{-1} (75 Wh dm^{-3}) may be obtained. Depending on use, service lifetimes of 3–5 years are normal.

For vehicles used in rugged terrain, batteries with tubular positive plates are required. As discussed above, SLI batteries are often supplied·in a dry charged state – for example in new cars.

Over the past ten years, the introduction of batteries described as 'maintenance free' (MF) has had an important impact on the SLI market. What this term implies in practice is that no addition of water to the electrolyte is required over a normal service life of 2–5 years. It has been

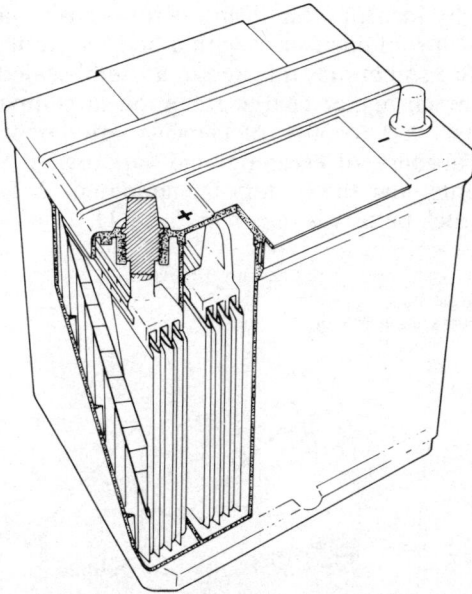

4.10 Cutaway diagram of typical SLI battery. (By courtesy of Magneti Mareili)

predicted that up to 80% of this market will be switched to MF units within the next few years.

Standard lead–acid batteries lose a small amount of water by evaporation, but the major mechanism for water loss is by electrolysis to form hydrogen and oxygen, as described by equations 4.5 and 4.6. The presence of small quantities of foreign elements lower the overvoltages for these processes and lead to an accelerated water loss: it has been shown that the elements antimony, arsenic, cobalt, manganese, nickel, platinum and tellurium all have a deleterious effect, even at low levels. Two general approaches have been taken in the development of MF batteries:

(i) reduction of the rate of gas formation within the normal operating conditions of the battery, and

(ii) promotion of gas recombination.

Since SLI batteries are usually recharged at constant voltage, MF versions must be constructed of such materials so that no substantial gassing occurs within the stabilised output voltage range of the alternator or dynamo. This implies that such batteries must contain modified positive grid and strap alloys in which the proportion of antimony is substantially reduced or eliminated and replaced by calcium or by a group of several metals (e.g. strontium/tin/aluminium). Calcium alloy grids have greatly improved behaviour so far as hydrogen evolution is concerned, but the calcium alloys have relatively poor processing characteristics – e.g. it is more difficult to fill complex grid moulds with such materials. MF batteries often have their electrolyte immobilised by means of gel formation using silica, calcium sulphate, etc., or by incorporating it in microporous separators. They are constructed in leak-proof enclosures with a one way vent to release excess gas pressure and are sometimes referred to as 'semi-sealed' systems. These batteries have a much higher charge retention in comparison with more conventional units – up to 30% of capacity after one year. A further advantage is the absence of escaping acid vapours, so that corrosion of metallic components near the battery is eliminated. A 12 V maintenance free semi-sealed truck battery is shown in Fig. 4.11.

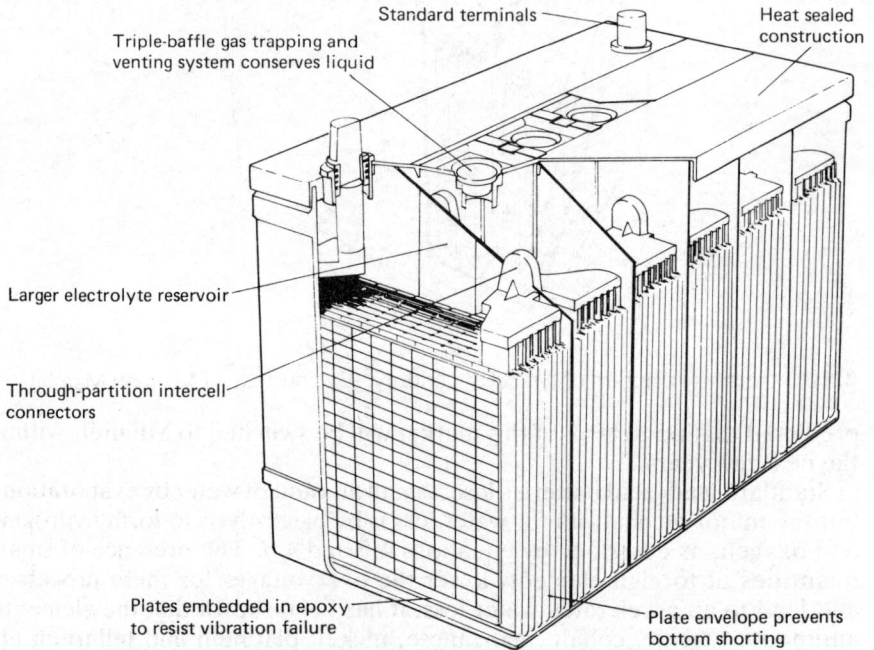

4.11   12 V maintenance free semi-sealed SLI truck battery. (By courtesy of Gould)

The production of hermetically sealed units requires the introduction of an efficient gas recombination process. In SLI batteries this may be accomplished by the incorporation of 'catalytic plugs' which promote hydrogen–oxygen recombination. However the feasibility of designing sealed MF units without such components has also been demonstrated: such 'recombining' cells rely on the reaction of lead with oxygen (and lead dioxide with hydrogen) and have particular design constraints. In charging, the cells are positive-limited, so that the oxygen evolved on overcharge must react with lead at the negative. To facilitate the transport of oxygen, the volume of the electrolyte may be restricted.

Industrial batteries

Motive power batteries are generally of higher quality than SLI batteries. Their most important characteristics are constant output voltage, high volumetric capacity at relatively low unit cost, good resistance to vibration and a long service life. Since electric motors used for motive power require high currents for long periods, traction batteries must be able to sustain prolonged and deep discharges followed by deep recharges, usually on a daily sequence. A typical discharge rate for this type of system might be C/5, and up to 80% of the nominal capacity would be required for daily service. The size and performance of traction batteries vary over a wide range: the voltage used may be 12–240 V and the capacity of each cell extends from 100 Ah to 1500 Ah or more. The specific energy of these units is normally in the range 20–30 Wh kg^{-1} (55–77 Wh dm^{-3}) and the cyle life is 1000–1500 cycles. In Fig. 4.12 charge and discharge curves for a typical traction battery are shown. Recently prototype lead–acid traction batteries with energy densities of 40 Wh kg^{-1} and 1000 cycle lifetime, and with energy densities up

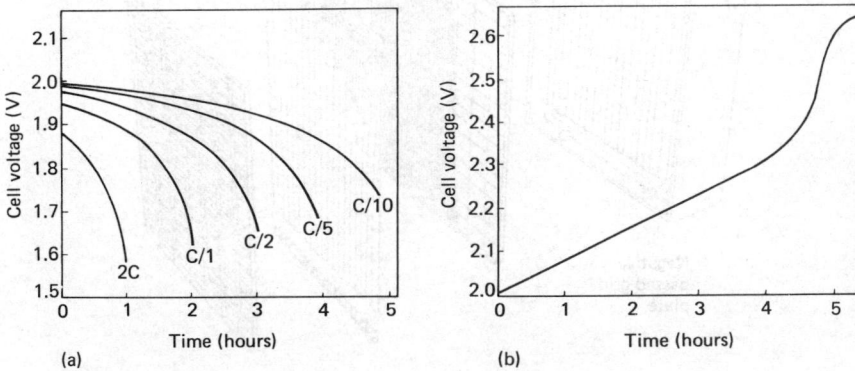

4.12 (a) Discharge curves for a typical lead–acid cell at various rates.
(b) Charging curve for the lead–acid cell at C/10.

to 60 Wh kg^{-1} but with lower cycle lives, have been announced by a number of manufacturers. In Fig. 4.13 the assembly of the type of cell used in a traction battery is shown: the multi-tubular positive plates give the cell high specific energy and capacity and assure a long cycle life. Positive plates are sometimes constructed using pasted grids, but in this case it is necessary to incorporate glass wool felt and special separators to absorb shocks and vibration, and to prevent shedding of active material. Traction batteries are used in industrial trucks, e.g. 'fork-lift' trucks (in the UK over 60% of all such vehicles are battery powered), milk floats and other delivery vehicles, mining and other tractors, industrial sweepers and scrubbers, golf carts, etc.

Stationary batteries fall into two groups: (i) systems for intermittent use or standby power and (ii) load levelling systems. The most important characteristics of the first group are reliability, long life and low self discharge rate. Such batteries often employ Planté plates; otherwise they use tubular plates or thick pasted plates formed on low antimony grids. An exploded view of a typical cell is shown in Fig. 4.14: the capacity of each cell is in the range 20–2200 Ah at C/10. The energy density is about half that of traction

Polypropylene container
and lid

Microporous "PORVIC®"
separator envelope

Positive tubular
plate

Negative
pasted·grid
plate

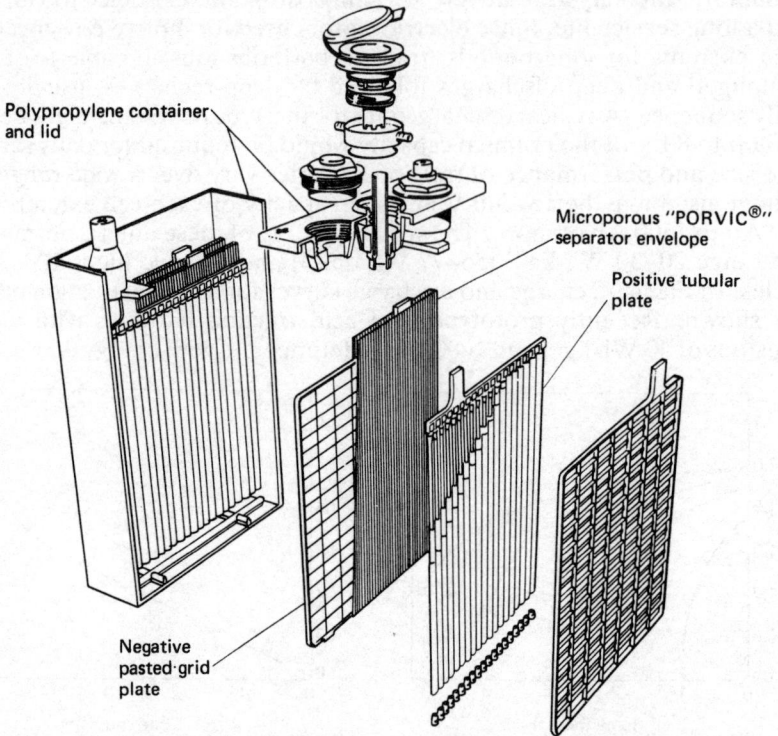

4.13   Motive power lead–acid cell with tubular positive plates in which the active material is contained in pre-formed terylene tubes, and negative pasted grid plates surrounded by micro-porous polyvinyl chloride separator envelopes. The case and lid are formed of heat sealed polypropylene. (By courtesy of Chloride Industrial Batteries)

batteries. Stationary batteries of this type are used for telecommunications systems, railroad signalling and track control and for standby power wherever it is necessary to have continuity of service during power cuts and momentary interruptions.

Over the past few years a number of studies have been made of the possible use of lead–acid batteries for load levelling. The service required is very similar to that of traction batteries except that energy density is less important than cycle efficiency.

·Sealed, portable batteries

In recent years the market for small, portable lead–acid batteries has grown considerably. These are sealed batteries obtained by assembling cylindrical or rectangular (prismatic) unit cells. They are generally not hermetically sealed since a safety vent is provided to allow gas escape in the event of excessive overcharge. The minimisation of water loss caused by electro-

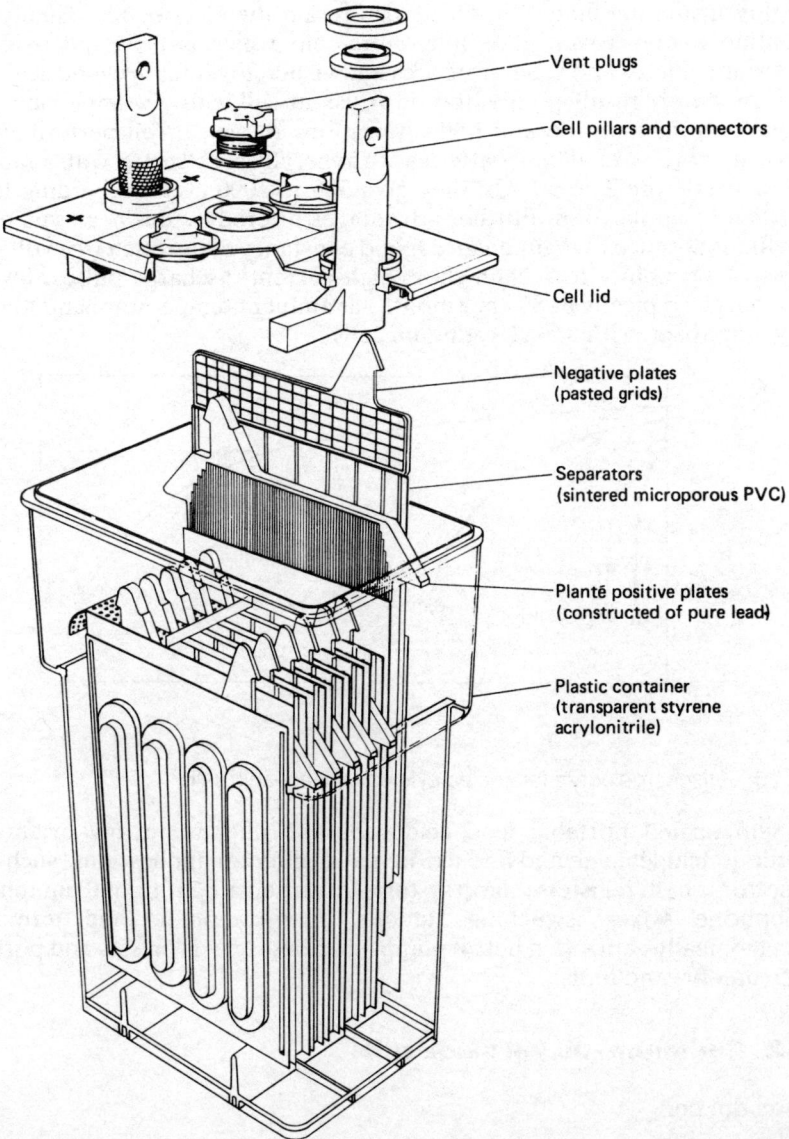

4.14 Standby power high performance Planté cell. The container is fabricated from transparent styrene acrylonitrile which enables the electrolyte level and cell condition to be easily monitored. (By courtesy of Chloride Industrial Batteries)

chemical dissociation into hydrogen and oxygen was considered above in connection with maintenance free SLI batteries.

Positive and negative plates are usually made using a 'honeycomb' grid support filled with active material. The separators are thin films of porous highly insulating materials which also retain the electrolyte. Usually they contain a non-woven glass microfibre mat which is heat and oxidation resistant. Plates and separators are sandwiched together in cylindrical rolls: this assembly results in a vibration resistant cell with low impedance, low polarization, long life and high utilization. A typical cell performance is shown in Fig. 4.15. These batteries are generally constructed with a range of capacities from 2 to 30 Ah; they give 300 to 2000 cycles according to the particular application. Further advantages offered by this new type of cell includes absence of maintenance, good constancy of the discharge voltage at up to C/4, ability to sustain short high current discharge pulses, low self discharge (typically 6–8% per month) at ambient temperature and low cost in comparison with nickel–cadmium cells.

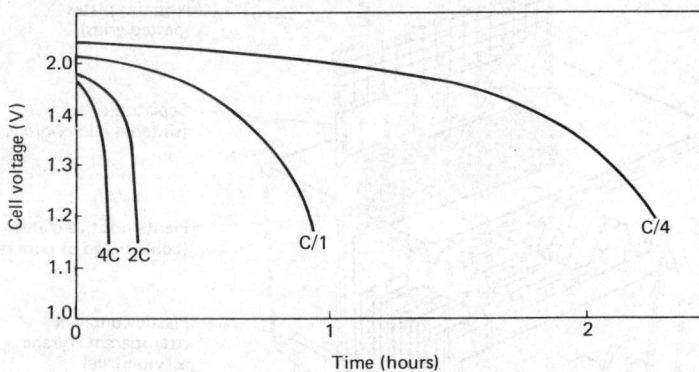

4.15   Discharge curve for sealed cylindrical lead–acid cell.

Semi-sealed portable lead–acid cells can operate in any orientation without acid leakage and find use in many different applications, such as in electronic cash registers, alarm systems, emergency lighting unit equipment, telephone boxes, switching stations, mini-computers and terminals, electronically controlled petrol pumps, cordless television sets and portable instruments and tools.

4.3 Cadmium—nickel oxide cells

Introduction

The first patent on an alkaline secondary battery was taken out by Waldemar Jungner of Sweden who in 1899 proposed a system based on nickel hydroxide as the positive electroactive material, a mixture of cadmium and iron as the negative electrode and an aqueous solution of potassium hydroxide as electrolyte. This cell, correctly called the cadmium–nickel oxide cell is almost universally referred to as the 'nickel–cadmium' cell and this latter term is used here. The nickel–cadmium system has been developed in different ways to produce a wide range of commercially important rechargeable systems including sealed maintenance free cells with capacities of 10 mAh–15 Ah, vented standby power units with capacities of over 1000 Ah, cranking batteries capable of delivering peak currents of 8000 A, etc. Nickel–cadmium cells are characterised by long life, continuous overcharge capability, relatively high rates of discharge and charge, almost constant discharge voltage and the ability to operate at low temperatures. However the cost of cadmium is several times that of lead, and this difference is unlikely to decrease since cadmium is a byproduct of zinc production. Moreover the cost of nickel–cadmium cell construction is generally more expensive than that of lead–acid cells so that the overall capital cost of energy storage is up to ten times higher. Health risks associated with the manipulation of cadmium may also affect future developments of this system. However, long cycle life, low maintenance and reliability have made it an obvious choice for a number of applications such as emergency lighting, electricity grid switching operations, engine starting, etc., while the sealed system which can operate in any orientation over a wide temperature range has found use in 'cordless' electrical appliances of many types – portable television receivers, hedge trimmers, electric shavers, etc. The good low temperature performance has led to the wide use of nickel–cadmium batteries in aircraft and space satellite power systems. At the present time, nickel–cadmium batteries account for more than 7% of total battery sales in the Western world market, and about 82% of all alkaline secondary battery sales.

The fully charged cell which may be written as

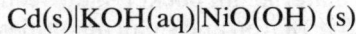

$$Cd(s)|KOH(aq)|NiO(OH) \ (s)$$

has an OCV of 1.30 V at ambient temperature and the basic overall cell reactions are

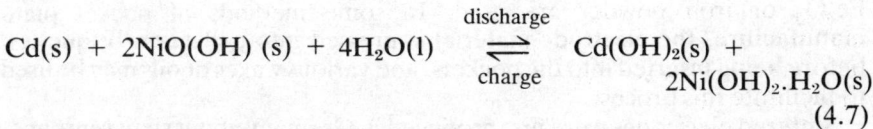

$$Cd(s) + 2NiO(OH) \ (s) + 4H_2O(l) \underset{charge}{\overset{discharge}{\rightleftharpoons}} Cd(OH)_2(s) + 2Ni(OH)_2.H_2O(s) \tag{4.7}$$

In practice, the cell reactions are more complex due to the formation of different NiO(OH) modifications and a series of higher nickel oxides with different degrees of hydration. The OCV of freshly charged cells may therefore be initially several hundreds of millivolts higher until these oxides spontaneously revert to NiO(OH) and oxygen. It is important to note that the electrolyte does not take part in the main cell reaction, so that its concentration is virtually independent of the state of charge of the cell: this is a particular advantage since neither the internal resistance nor the freezing point of the electrolyte are affected by the state of charge. (A minor disadvantage is that there is no simple way of assessing the state of charge, as can be done with the lead–acid battery by measuring the electrolyte density.) Depending on construction, nickel–cadmium cells have practical energy densities in the range 10–35 Wh kg^{-1} (30–80 Wh dm^{-3}) and a cycle life ranging from several hundreds for sealed cells to several thousands for vented cells.

Cell construction is mainly confined to two types, using either 'pocket plate' electrodes (vented cells) or 'sintered plate' electrodes (vented and sealed cells). In the former, the active materials are retained within pockets of finely perforated nickel–plated sheet steel which are interlocked to form a plate. Positive and negative plates are then interleaved with insulating spacers placed between them. In sintered plate electrodes, a porous sintered nickel mass is formed and the active materials are distributed within the pores. In sintered plate vented cells, cellulose or other membrane materials are used in combination with a woven nylon separator. In sealed or 'recombining' cells, special nylon separators are used which permit rapid oxygen diffusion through the electrolyte layer.

Negative electrodes

The electrochemical reactions at the negative electrode are the comparatively straightforward processes

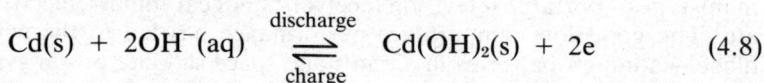

$$Cd(s) + 2OH^-(aq) \underset{charge}{\overset{discharge}{\rightleftharpoons}} Cd(OH)_2(s) + 2e \qquad (4.8)$$

However it is recognised that slightly soluble intermediates such as $CdO(OH)^-$ and $Cd(OH)_3^-$ are involved. Cadmium does not corrode since its equilibrium potential is more positive than that of hydrogen in the same solution. The active material in pocket plate cells consists of metallic cadmium, with up to 25% of iron and small quantities of nickel and graphite to prevent agglomeration. Two methods of preparation are used. One involves the electrochemical co-reduction of a solution of cadmium and iron sulphate; in the other, dry mixtures of cadmium oxide or hydroxide and Fe_3O_4 or iron powder are used. In some methods of pocket plate manufacture, the electrode material is pressed into pellets or 'briquettes' before being inserted into the pockets, and various waxes or oils may be used to facilitate this process.

Sintered electrodes were first produced in Germany about sixty years ago, but large scale commercial manufacture did not start until after the Second World War. The preparation of sintered electrodes is more difficult and more costly than that of pocket plate electrodes, but leads to cells with superior electrical and mechanical characteristics. For high rate applications, accurate, thin section, closely spaced electrodes can be produced and since the active material is in close contact with the current collector, the cell resistance remains fairly constant right up to the end of the discharge. The sintered plate substrates are made by pressing and then sintering at 800–1000°C in hydrogen nickel powder (usually formed by thermal decomposition of $Ni(CO)_5$) on a supporting nickel perforated foil or wire screen. By incorporating gas forming materials such as $(NH_4)_2CO_3$ in the nickel powder before sintering, a highly porous, mechanically stable matrix is produced. Up to 75–85% of the volume is void, and by carefully controlling the starting materials and thermal treatment, uniform pore radii ($\approx$5–30 μm) can be guaranteed.

The sintered product is cut into the required shape and the active material is infused using one of a number of techniques – e.g. by impregnating the sinter with concentrated aqueous cadmium nitrate, followed by thermal decomposition, or by cathodic polarisation in molten $Cd(NO_3)_2$ baths, etc. The plates are washed and the impregnation cycle repeated up to 5–10 times until the required loading is attained. Finally the plates are 'formed' by a sequence of carefully controlled charge/discharge cycles. Safety precautions are very important in the manufacture of cadmium-based electrodes because of the health hazards associated with this material.

Positive electrodes

The principal electrochemical reactions are represented by the equation:

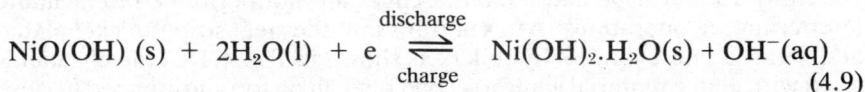

$$NiO(OH)\ (s)\ +\ 2H_2O(l)\ +\ e\ \underset{charge}{\overset{discharge}{\rightleftharpoons}}\ Ni(OH)_2.H_2O(s) + OH^-(aq)$$

$$(4.9)$$

However, as pointed out above, some hydrated higher oxides of nickel (up to $NiO_{1.8}$) are initially formed on charge, which slowly decompose to give $NiO(OH)$ and oxygen. This situation is further complicated by the involvement of KOH (and possibly LiOH) from the electrolyte, and the formation of mixed alkali metal–nickel oxides. The net result of these additional reactions is an elevated OCV in freshly charged cells and some loss of cycle efficiency. Depending on charging rate and other factors, different crystal modifications of $NiO(OH)$ in addition to the predominant β–form may be produced, some of which cause considerable swelling and possible deterioration of contact between the active material and the current collector.

The active material used in pocket plate cells consists of $Ni(OH)_2$ together with up to 5% of $Co(OH)_2$, $Ba(OH)_2$ etc. to improve cell capacity and cycle life, and 20% of graphite in various forms to increase the electronic conductivity. The nickel hydroxide is precipitated from nickel sulphate in a controlled manner to produce fine particles of large surface area. As in the case of the cadmium electrode, the nickel hydroxide powder may be formed into pellets before insertion into the pockets.

Sintered electrodes are prepared in exactly the same way as for the negative electrodes.

Electrolyte

The electrolyte generally used is an aqueous solution of potassium hydroxide with a concentration of 20–28% by weight and a density of 1.18–1.27 g cm^{-3} at 25 °C. The low end of the range is used for cells designed to operate at normal temperatures (say above -10 °C). One to two per cent of lithium hydroxide is usually added to such electrolytes to minimise co-agulation of the $NiO(OH)$ electrode on cycling. For low temperature applications, the more concentrated KOH solutions are used (without LiOH which increases electrolyte resistance); some problems with swelling are encountered with pocket plate electrodes and these concentrated solutions.

Cells operating at high temperatures occasionally use aqueous sodium hydroxide as electrolyte.

Some water loss occurs, especially if there is prolonged overcharging of cells. Cells may be designed so as to contain a large reserve of electrolyte, so that topping up is only necessary after long intervals – e.g. 5–8 years in some stationary standby applications.

Vented pocket plate cells

The pocket plates are made from low carbon mild steel strip, 0.1 mm thick and perforated with 250–500 circular holes per square centimetre by needle or roller perforators. Each hole has a radius of about 0.1 mm. The strip is generally 16 mm wide and the outer edges are not perforated to facilitate later crimping operations. After perforation, the steel strip is nickel plated and formed into channels by rollers as shown in Fig. 4.16. One channel is filled with active material while a second is fed in on top and crimped to close the pocket. The required number of filled strips is then interlocked together to form a battery plate. After guillotining to size, the cut edges are enclosed

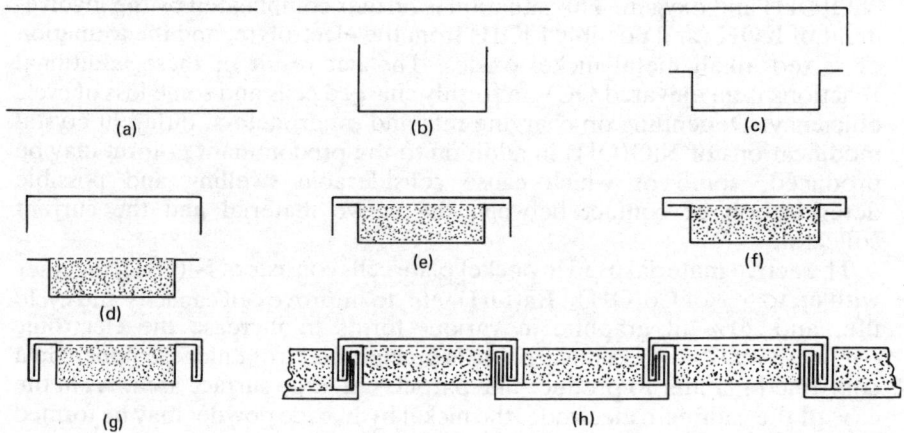

4.16 Sequence of operations in fabrication of pocket plates for nickel–cadmium batteries.

    (a)–(c):   Formation of channels in nickel-plated perforated steel strip.
    (d)–(f):   Filling and crimping to form a long continuous pocket.
    (g)–(h):   Interlacing of filled strips and compression to form final plate.

by welded U-section plated steel strip and current collector lugs are attached by spot welding. Finally the plates are pressed to form a flat rigid structure with a corrugated surface which permits insulating separator rods to be inserted (see Fig. 4.17).

Plates of the same polarity are then assembled with spacing washers on a collector bar to which is also attached the terminal pillar. Large assemblies are usually bolted together, while in small capacity systems the plates may be welded together. Double positive plates are used for heavy duty batteries

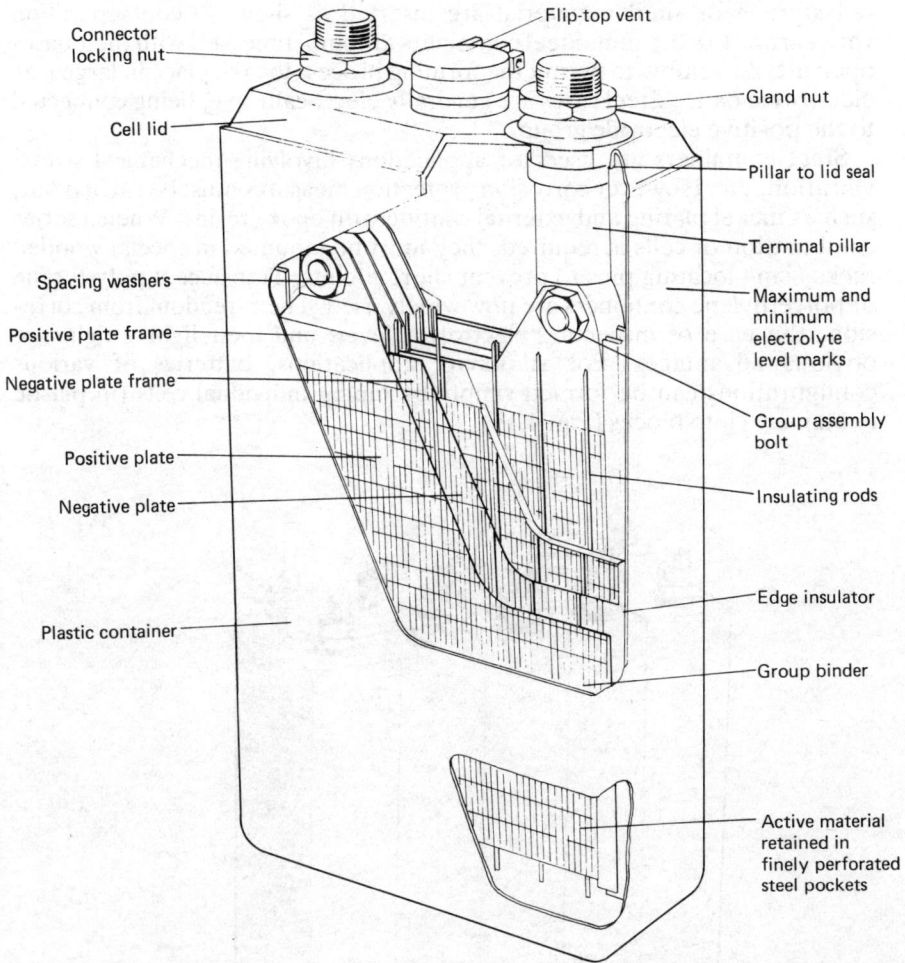

4.17 Cutaway diagram of a typical nickel–cadmium cell. (By courtesy of Chloride Alcad)

with deep cycling regimes, while thinner plates are used in batteries designed for high rate/short discharge applications. Negative and positive electrode groups are interleaved as seen in Fig. 4.17, so that plates of opposite polarity are adjacent to one another and the thin insulating rods of polystyrene or similar material are inserted as shown. Plate separation varies from 1.0–3.5 mm. Steel containers are sometimes left with their bases open after assembly to permit the forming process to take place in large free electrolyte baths. Steel cases are generally electrically live, being connected to the positive electrode group.

Steel containers are used for applications involving mechanical stress, vibration, etc. However corrosion protection measures must be carried out, such as nickel plating and external coating with epoxy resins. When a series combination of cells is required, they must be mounted in special wooden racks using locating pins to prevent short circuits. Translucent polystyrene or polyethylene containers are now widely used. Their freedom from corrosion, the ease of inspecting electrolyte levels and their light weight are obvious advantages. For stationary applications, batteries of various configurations can be formed simply by taping individual cells (in plastic containers) into blocks (Fig. 4.18).

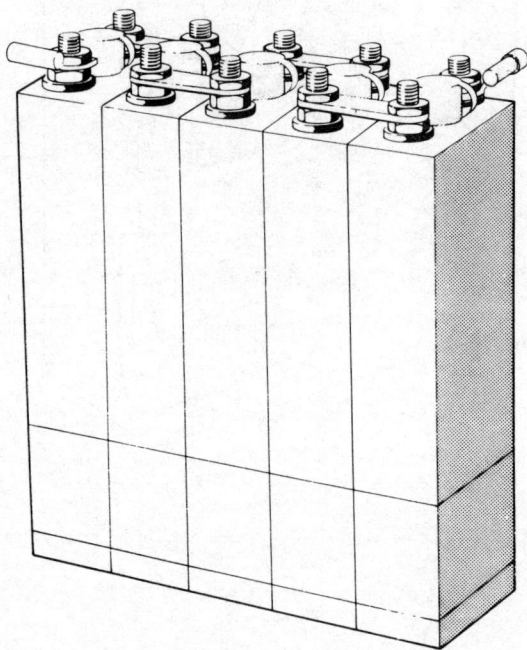

4.18 Nominal 6 V battery formed by connecting five nickel–cadmium cells in series.

Vented sintered plate cells

Sintered plate cells are assembled in a similar manner. Since the products of charging are formed within the pores of the rigid electrode matrix, there is no chance of swelling and the plates can be mounted closer together. Most manufacturers use woven nylon or other synthetic fibre separators; perforated polyvinylchloride sheet is also used.

Vented cells with tubular electrodes

Some versions of the cell are manufactured in which the active materials are supported in tubular elements made from spirally wound perforated steel ribbon. Such a design is more common for alkaline nickel–iron cells, and will therefore be described for the latter system.

Performance

Pocket plate cells have energy densities in the range 10–25 Wh kg^{-1}, depending on whether the cell has been designed for low or high rate duties. Sintered plate cells have a 50% higher energy density and can maintain higher rate discharges since they have lower internal resistance. Recently increased power and energy densities have been achieved in 40 Ah aircraft batteries by optimising design parameters and carefully choosing the separator material. Batteries of this type (24 V) can supply up to 23 kW of instant power at 25 °C with a power density of 600 W kg^{-1}. At -30 °C the power is reduced by only 50%. Standing losses are lower for pocket plate than for sintered electrode cells. The inherent long life of the former (25–30 years in stationary batteries) is a notable advantage for emergency power applications.

Typical charge/discharge curves for nickel pocket plate cells at a number of rates are shown in Fig. 4.19. The relatively flat character of the discharge curves at up to the 5 hour rate is noteworthy. The distinct rise in the charging potential at about 90% of the capacity is due to the changeover from cadmium ion reduction to hydrogen evolution. Since some oxygen evolution occurs on the positive electrode during charge, it is necessary to supply up to

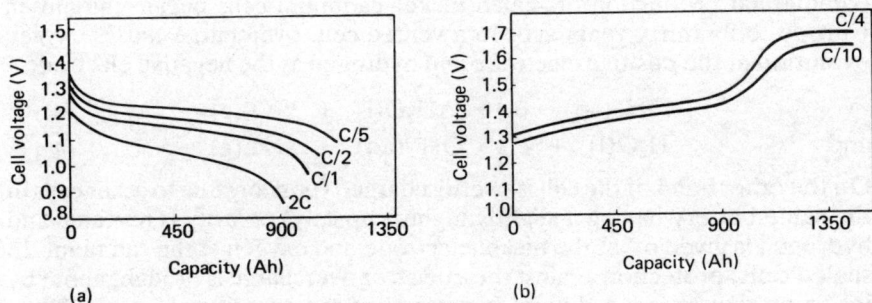

4.19 Discharge (a) and charge (b) characteristics of a typical 900 Ah nickel–cadmium battery as a function of rate.

25% excess charge for sintered plate cells and 50% for pocket plate cells in order to achieve complete conversion of $Ni(OH)_2$: hence the cycle energy efficiencies of the two types of cell are 68% and 55% respectively. After an initial fairly rapid self-discharge (up to 20%, say), self-discharge of nickel–cadmium cells is very slow, except at elevated temperatures. In Fig. 4.20 charge retention is shown for a typical pocket plate cell as a function of time and temperature. It is seen that at 25 °C, 80% of the capacity is still available after 12 months. Self-discharge is higher with sintered plate electrodes. Unlike the lead–acid cell, nickel–cadmium cells can be stored for extended periods at any state of charge without damage.

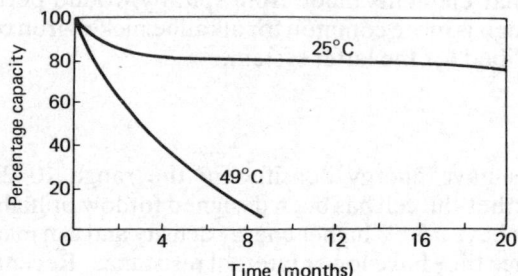

4.20   Charge retention in nickel–cadmium cells after prolonged periods of open circuit. (By courtesy of Chloride Alcad)

Electrolytes may be contaminated by carbon dioxide over a number of years. When the carbonate concentration reaches a value of over 60 g dm^{-3} the electrolyte must be replaced.

Sealed cells

Sealed nickel–cadmium secondary cells are designed so that no significant build-up of gas pressure occurs under normal working conditions, and since the electrolyte remains invariant, they require no maintenance. Edison patented an alkaline iron–nickel oxide sealed cell system in 1912, but commercial production of sealed nickel–cadmium cells began, initially in Europe, only thirty years ago. In a vented cell, overcharge causes oxygen evolution at the positive electrode and hydrogen at the negative electrode:

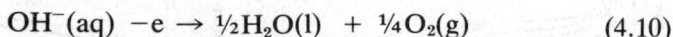

$$OH^-(aq)\ -e \rightarrow \tfrac{1}{2}H_2O(l)\ +\ \tfrac{1}{4}O_2(g) \qquad (4.10)$$
and
$$H_2O(l)\ +\ e \rightarrow OH^-(aq)\ +\ \tfrac{1}{2}H_2(g) \qquad (4.11)$$

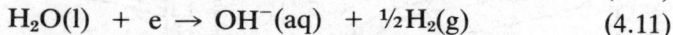

On the other hand, if the cell is overdischarged (possibly due to other cells in the same battery having a slightly higher capacity) polarity is reversed and hydrogen is given off at the nickel electrode and oxygen at the cadmium. In sealed cells, protection against the effects of overcharge is brought about by incorporating excess cadmium hydroxide in the negative electrode. Then when the positive electrode becomes fully charged, the negative electrode is still only partially charged. A continuation of the charging current thus results in oxygen evolution at the positive electrode (by the reaction of equation 4.10), but further cadmium hydroxide reduction at the negative. The free oxygen can now diffuse to the negative electrode where it may be reduced electrochemically or react with cadmium. The cycle of oxygen evolution at the positive and consumption at the negative electrode can continue indefinitely without affecting the cell. Cells can be fabricated which allow overcharge at C/10 without oxygen pressure ever exceeding one atmosphere. Note too that the water content of the electrolyte remains constant. On terminating the overcharge current oxygen pressure falls as the oxygen continues to be consumed by

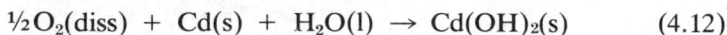

$$\tfrac{1}{2}O_2(\text{diss}) + Cd(s) + H_2O(l) \rightarrow Cd(OH)_2(s) \qquad (4.12)$$

Protection against overdischarge is achieved by the incorporation of a quantity of cadmium hydroxide – known as the 'antipolar mass' – in the positive electrode. Overdischarge then results in the formation of cadmium by reduction of the antipolar mass, rather than the evolution of hydrogen. Any oxygen evolved at the exhausted cadmium electrode will diffuse to the positive electrode and combine with the cadmium according to equation 4.12. It should be noted that the recombination reaction is exothermic, so that heat generation on overcharge can be a serious problem, especially with larger cells which, by their nature, have a relatively lower surface/volume ratio.

Sealed nickel–cadmium cells are manufactured as button, cylindrical and rectangular (prismatic) cells with nominal capacities ranging from 10 mAh to 15 Ah. The electrodes are commonly of the sintered type and their preparation is similar to that described above for vented cells. Some manufacturers incorporate polymeric bonding materials and use special pressing techniques. Button cells may also be assembled with circular pocket electrodes formed by pressing the active materials and containing the resulting pellets in fine nickel mesh. A separator disc of several layers of non-woven nylon or cellulose is inserted between the electrodes. Separator material and design is important in sealed cells, since oxygen flow between the electrodes must not be greatly impeded. A typical button cell of this type is shown in Fig. 4.21. Generally the positive electrode is placed at the bottom of the case in contact with an expanded metallic spacer. The negative

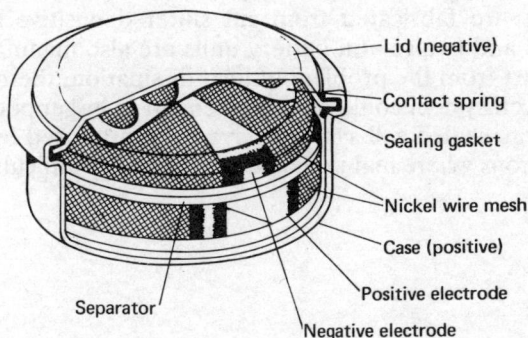

4.21 Sealed nickel–cadmium button cell. (By courtesy of Varta)

electrode makes contact with the lid by means of a steel spring which serves to ensure good contact between the electrode masses and the electrolyte containing separator. Button cells are used individually, or as series pack assemblies with different operating voltages, as illustrated in Fig. 4.22.

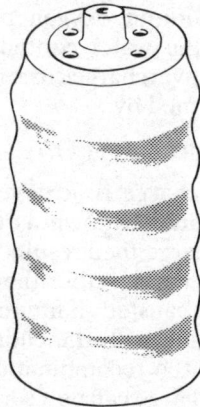

4.22 Nominal 6 V sealed nickel–cadmium battery formed by connecting five cells in series. (By courtesy of Varta)

Special versions of these, usually with a nominal 3.6 V output and having appropriately sited terminals, are produced for mounting on printed circuit boards. Cylindrical cells, often manufactured as replacements for primary Leclanché cells of similar dimensions, are manufactured in two designs. In the most common version, a spiral construction is used, as shown in Fig. 4.23. The case is constructed of nickel plated steel and acts as the negative terminal. The insulating disc at the bottom of the cell prevents short circuits between the rolled electrodes and the case. Note that the majority of cylindrical and rectangular cells have safety valves to release any excessive internal pressure resulting from, say, uncontrolled overcharging. Cylindrical cells with higher specific capacity but lower power have a bobbin-type construction with the pressed positive electrode at the centre. Rectangular (prismatic) cells are fabricated from cut sintered positive and negative plates. Six, nine and twelve volt battery units are also manufactured using this design. Apart from the problem of heat dissipation, there is no reason why the sealed cell system could not be extended to larger pocket cells and indeed some large sealed cell systems have been produced for emergency lighting applications where maintenance is particularly difficult.

Applications

(i) Vented cells

Vented pocket plate cells have a wide range of applications that can be subdivided into three main groups. In the first, the cells are normally

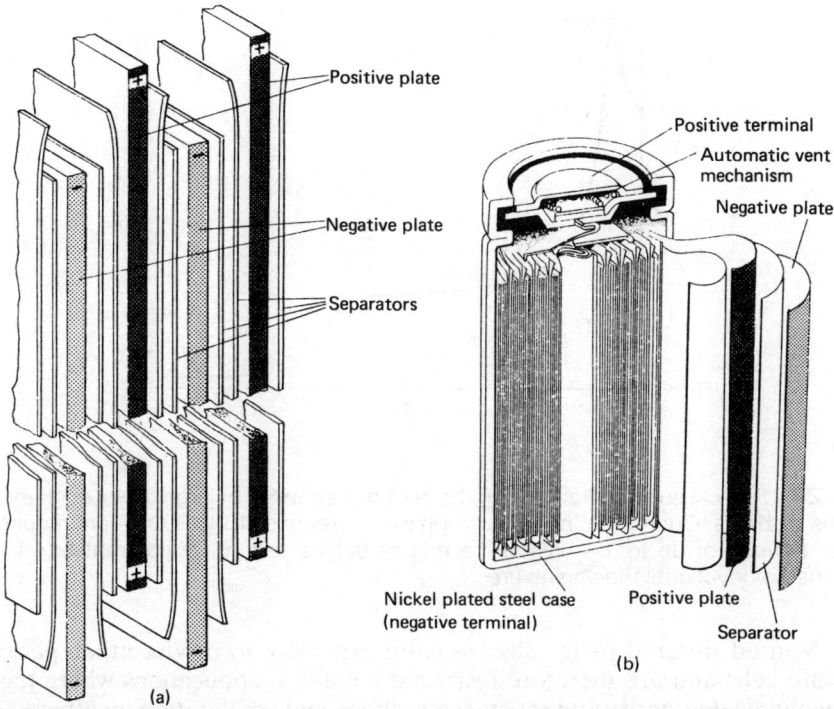

4.23 Schematic cross-section of spiral wound cylindrical sealed nickel-cadmium cell.

subjected to only shallow discharge, with occasional deep discharges and are kept on a floating charge. The long life (> 30 years) and low maintenance requirement of nickel–cadmium cells under such a regime make them very suitable for emergency lighting, switch tripping in the electricity distribution industry, electric train control duties, etc. In the second group, cells are subjected to deep cycling at moderate rates (e.g. C/1–C/3) in such applications as train lighting, marine duties, telecommunications and traction (mine locomotives, industrial trucks, etc.). The third group comprises high rate cells with some deep cycling capability. Their main use, apart from short period emergency lighting, is for starting large diesel engines and gas turbines. A typical current demand curve for engine starting is shown in Fig. 4.24. In the initial 'breakaway' phase, currents of 1000–8000 A are required for periods of up to one second. Cells with large numbers of thin plates can deliver currents of greater than 10 C for this period, with the cell voltage dropping to no less than 0.65 V. The 'cranking current', usually about half that of the breakaway, may be needed for periods of 5–30 s before the engine starts, and a typical specification requires the starting sequence to be repeated six times. (Voltage recovery is almost instantaneous in nickel–cadmium cells, unlike the situation for the lead–acid system.) In most cases, except for very low temperature operations, the cranking current requirement determines the battery size.

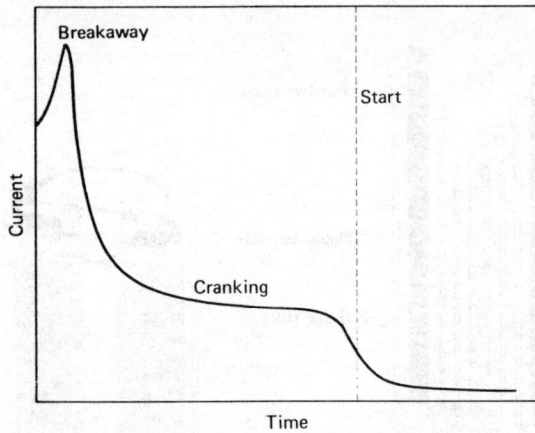

4.24 Typical current drain during the starting sequence of large diesel engines or gas turbines. During the 'breakaway' phase, currents of 1000-8000 A are required for periods of up to 1 s. When the engine begins to turn, the current falls to a constant level until the engine fires.

Vented sintered plate cells are more expensive to produce than pocket plate cells and are therefore restricted mainly to applications where their mechanical integrity and resistance to shock and acceleration together with their high power density are of importance – e.g. in aircraft, helicopters, military vehicles, city buses, etc. 24 V batteries are generally used, with capacities in the range 30–60 Ah.

(ii) Sealed cells

Nickel–cadmium sealed cells are now a commercially important consumer product. Both as button and cylindrical cells, they find use in portable 'cordless' appliances such as power tools, electric razors, photoflash apparatus and increasingly in 'hybrid' mains/battery equipment such as portable tape recorders, radios and television receivers. Many of these cells are readily interchangeable with primary batteries. In recent years advances in design have increased recharge rates: cylindrical cells with sintered electrodes can now be fast charged from full discharge at up to the C/1 rate to 80% of capacity.

Sealed cells have also many important military and aerospace applications where absence of maintenance may be important. The battery for the 'Viking' Mars orbiting spacecraft consisting of twenty six sealed 30 Ah cells, is shown in Fig. 4.25. This spacecraft was placed in Mars orbit in 1976.

Nominal 3.6 V batteries have been designed for direct mounting on printed circuit boards for CMOS and NMOS memory support applications. Such cells are normally float charged from the main microcomputer d.c. power supply and constitute a form of uninterruptable power supply (UPS).

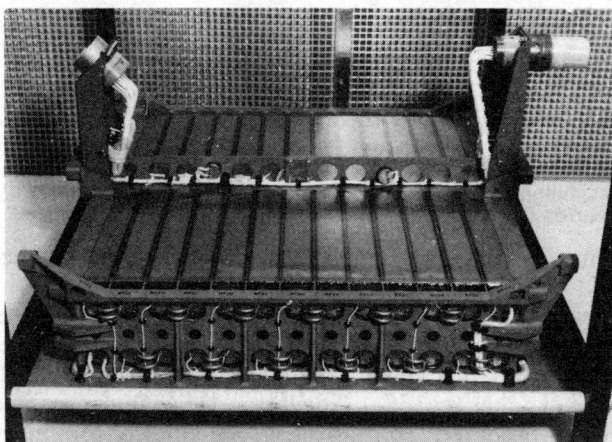

4.25 Battery for 'Viking' Mars orbiting spacecraft, comprised of 26 sealed 30 Ah nickel–cadmium cells, which was placed in Mars orbit in 1976. (By courtesy of Jet Propulsion Laboratory)

4.4 Iron–nickel oxide cells

Introduction

The iron–nickel oxide alkaline battery system has many features in common with the nickel–cadmium system discussed above. It was first developed by Edison in the USA at the turn of the century and was patented in the same year as Jungner's first nickel–cadmium US Patent, 1901. Iron can be re-garded as a favourable active battery material because of its low cost, high theoretical specific capacity (twice that of cadmium) and non-toxic, pollution-free characteristics. However because its reduction potential is below that of hydrogen and since hydrogen overvoltage is low on iron, charge retention is poor, and efficiency is low.

Manufacture of iron–nickel oxide batteries commenced in 1908, but the system did not have the commercial success of nickel–cadmium. Until recently there was only a very limited production of stationary batteries in the USA, Germany and the USSR. Developments of improved iron elec-trodes have altered the situation, and the iron–nickel oxide system is now being actively considered for EV propulsion and other applications.

The cell in fully charged state can be written as

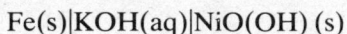

$$Fe(s)|KOH(aq)|NiO(OH) (s)$$

and has an OCV of about 1.41 V at 25 °C (higher when freshly charged due to the presence of higher oxides of nickel). The basic cell reactions are

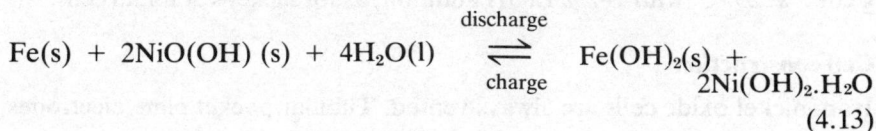

$$Fe(s) + 2NiO(OH) (s) + 4H_2O(l) \underset{charge}{\overset{discharge}{\rightleftharpoons}} Fe(OH)_2(s) + 2Ni(OH)_2.H_2O$$

$$(4.13)$$

It is probable that a range of soluble species such as $Fe(OH)_2^-$ and FeO_2^- are involved and it is known that $Fe(OH)_3$ or Fe_3O_4 may be formed on deep discharge. The practical energy density of conventional tubular plate cells is $20-30\,Wh\,kg^{-1}$; with the more recent cells which use press-sintered iron electrodes, values of $40-60\,Wh\,kg^{-1}$ have been reported.

Positive electrodes

The most common configuration is of vertical rows of tubular pocket electrodes held in a nickel plated steel frame. The tubes are manufactured by spirally winding a perforated plated steel ribbon. The tubes are re-inforced by plated steel rings placed at fixed distances along their length, and are packed with alternating layers (> 30 per cm) of dried nickel hydroxide powder and nickel flakes ($\approx 13\%$). The latter, which improve the electronic conductance, are prepared from an electrochemically formed multilayer sheet of copper and nickel which is cut and then immersed in sulphuric acid to dissolve out the copper. The tubes are closed and pressure-inserted or welded into the frame to form a unit similar in construction to the $Pb-PbO_2$ tubular system described in section 4.2. These electrodes are expensive and difficult to manufacture, but they are very rugged and can readily withstand the stress caused by expansion of the active material. Their life can exceed seven years, even with heavy cycling duty.

Recently, cells employing thick sintered nickel plates on nickel plated porous steel substrates have been developed which have greatly improved energy densities. The active material is introduced by electroprecipitation. Electrodes based on nickel fibre supports are also being studied.

Negative electrodes

In pocket plate cells, the active materials are a mixture of finely powdered metallic iron and Fe_3O_4. The preparation of this mixture varies from manufacturer to manufacturer, but generally involves a final process in which controlled air oxidation of iron powder or reduction of Fe_3O_4 with hydrogen is used to form the appropriate composition. Additives such as cadmium, cadmium oxide or graphite are commonly included to improve the capacity retention and electronic conductance. The performance of the electrode is improved by the addition of up to 0.5% of FeS; the mechanism of the sulphide involvement is not well understood. If sulphide is lost by oxidation after prolonged use, small amounts of soluble sulphide may be added to the electrolyte.

Electrolyte

The electrolyte is aqueous KOH with a density of approximately $1.22-1.30$ $g\,cm^{-3}$ at $25\,°C$, with $1-2\%$ LiOH addition, as for nickel–cadmium cells.

Cell construction

Iron–nickel oxide cells are always vented. Tubular/pocket plate electrodes are constructed as described above and are generally housed in nickel plated steel cases. Cells with sintered plate electrodes have smaller inter-electrode spacings. They use synthetic fibre fabrics as separators, and plastic containers.

Performance and applications

A typical charge/discharge curve at ambient temperatures at the C/3 rate is shown in Fig. 4.26. A mean discharge voltage of about 1.2 V is realised. Commercial cells normally have capacities in the 250–600 Ah range. Cycle efficiencies are rather low because of the need for extensive overcharge. Low temperature performance is poor compared with that of nickel–cadmium cells, but the most important handicap of this system is its high rate of self-discharge: on one months' storage at ambient temperature, 30–50% capacity is lost.

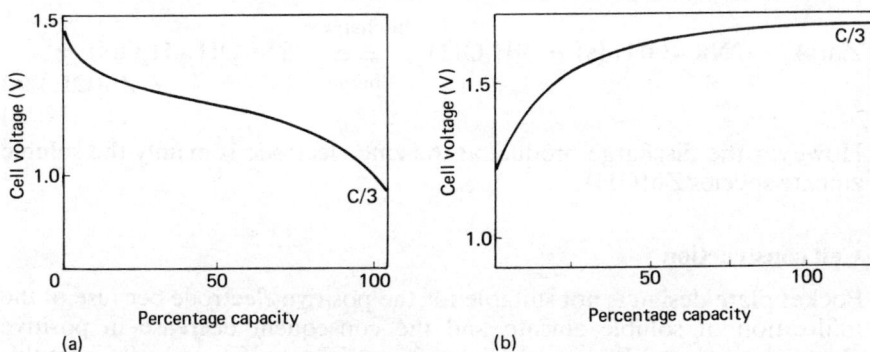

4.26 Discharge (a) and charge (b) characteristics of a typical iron–nickel oxide cell at C/3, as a function of percentage capacity.

The theoretical energy density of the cell is $268 \, Wh \, kg^{-1}$. Practical values of $40–60 \, Wh \, kg^{-1}$ are common in EV propulsion battery prototypes, and Matsushita have claimed $82.5 \, Wh \, kg^{-1}$ at C/5. Electric cars fitted with the latter system have achieved 200 km for a single charge, and a cycle life of 1500–2000 has been attained.

Iron–nickel oxide batteries have been used for many years in railway lighting applications, for motive power in industrial trucks, tractors and mine locomotives. There is some utilisation of the system in emergency lighting and alarm circuits, and there has been renewed interest recently when a number of companies have begun developing EV batteries.

4.5 Zinc–nickel oxide cells

Introduction

Secondary battery systems with zinc electrodes in alkaline solution have been in existence for a hundred years, and in the 1930's the Drum railway battery, an early zinc–nickel oxide power source was used for railway traction on a regular passenger line in Ireland. However problems associated with the zinc electrode reduced commercial interest in the system so that ten years ago no zinc–nickel oxide battery was being manufactured. Yet many now consider this system to be the most promising short term solution to the problem of power supply for EV propulsion. The zinc–nickel oxide cell has a number of important advantages, in particular its high operating voltage and energy density, good high rate discharge characteristics and low cost. The theoretical specific capacity of zinc ($825\ Ah\ kg^{-1}$) is almost as high as that of iron and over three times that of lead. The main disadvantage of the system is its relatively poor cycle life and charge retention.

The charged cell, which may be represented as

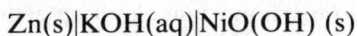

$$Zn(s)|KOH(aq)|NiO(OH)\ (s)$$

has an OCV of 1.78 V at 25 °C (higher when freshly charged due to the presence of higher oxides of nickel). The cell reactions are formally:

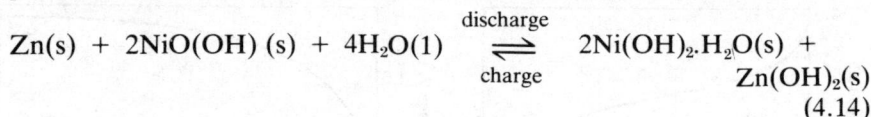

$$Zn(s) + 2NiO(OH)\ (s) + 4H_2O(1) \overset{\text{discharge}}{\underset{\text{charge}}{\rightleftharpoons}} 2Ni(OH)_2.H_2O(s) + Zn(OH)_2(s)$$

$$(4.14)$$

However the discharge product at the zinc electrode is mainly the soluble zincate species $Zn(OH)_4^{2-}$.

Cell construction

Pocket plate design is not suitable for the positive electrode because of the infiltration of soluble zincate and the consequent decrease in positive electrode capacity. Porous matrix positives do not suffer so badly from this problem. Conventional sintered plate positives are relatively heavy and expensive because of the quantity of nickel required as support material. For zinc–nickel oxide EV cells new techniques have been developed to produce a polymer-bonded matrix using rolling or pressing methods. Graphite is commonly added to improve the electronic conductance and cobalt to improve capacity retention. Positive electrodes based on nickel fibres have also been studied.

The recharging of zinc electrodes has been widely studied as the development of zinc–air, zinc–silver oxide and zinc–nickel oxide secondary batteries has progressed. The main problems concern the morphology and distribution of the electrodeposited zinc. Under certain conditions zinc is deposited as dendrites which can penetrate separators and short circuit the cell. Alternatively poorly adherent deposits are formed so that the capacity of the electrode is progressively reduced. Further, 'shape changes' may occur as zinc is stripped from the electrode edges during discharge and redeposited near the electrode centre during charge.

To minimise the chance of dendritic growth it is important to ensure that the zincate ion concentration never falls to very low values. This is accomplished by making the cells nickel limited. Since NiO(OH) electrodes require 20–30% overcharge for full capacity (due to oxygen evolution), it is general practice to make the capacity of the zinc electrode at least twice that of the positive electrode. In sealed cells, oxygen evolved at the positive recombines with zinc at the negative electrode or is electrochemically reduced so that zincate ion concentration is not depleted even on prolonged overcharge. However severe overcharging in vented cells can result in over-deposition of zinc: under these circumstances the cell must be short-circuited to restore the zincate concentration.

Investigations are being made into alternative zinc electrode configurations. Contoured electrodes containing PTFE powder have been found to extend cell life and the use of surface-active agents has been reported to be successful.

Agitation of the electrolyte by a rotating sector or mechanical vibration of the zinc during charge leads to the deposition of more adherent compact layers with a reduction in shape and dendrite problems.

One of the most vital aspects of cell design in the zinc–nickel oxide system is the development of durable separators. Cellulose materials swell in aqueous KOH, causing a tight packing in the cell which helps to preserve the integrity and shape of the zinc electrode. However oxygen evolved from the positive electrode on charge can cause degradation of cellulose leading to production of CO_2 and carbonation of the electrolyte. A range of alternative organic and microporous inorganic separators is being studied.

Performance and applications

Charge/discharge cycle characteristics for a cell operating at room temperature are shown in Fig. 4.27. The high cell voltage under load (>1.5 V even at a rate of 2C) is noteworthy.

Almost all the development of this system is directed towards EV applications. Prototype full scale batteries have been tested by a number of companies in the USA, Europe and Israel. Energy densities of up to 73 Wh kg^{-1} (155 Wh dm^{-3}) can be produced by 300 Ah cells and now 90 Wh kg^{-1} (190 Wh dm^{-3}) systems are being developed. Road tests with small passenger vehicles have achieved ranges of 100–230 km on a single charge.

Prototype button cells have been fabricated and studied.

4.6 Zinc–silver oxide and related cells

Introduction

The zinc–silver oxide couple has been known for over a hundred years and forms one of the highest energy aqueous cells. The theoretical energy density is 300 Wh kg^{-1} (1400 Wh dm^{-3}) and practical values in the range 40–110 Wh kg^{-1} (116–320 Wh dm^{-3}) have been achieved. Primary cells were considered briefly in section 3.6.

The first successful secondary zinc–silver oxide alkaline battery system was developed in France by André in the late 1930's. He overcame the problem of migration of silver (as $Ag(OH)_2^-$, $Ag(OH)_3^{2-}$ and similar

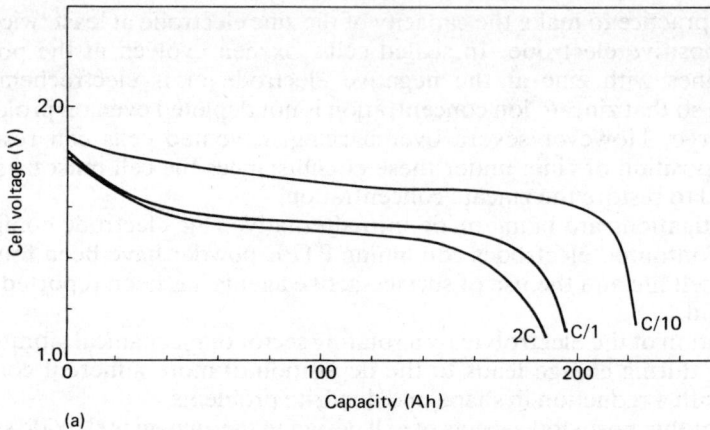

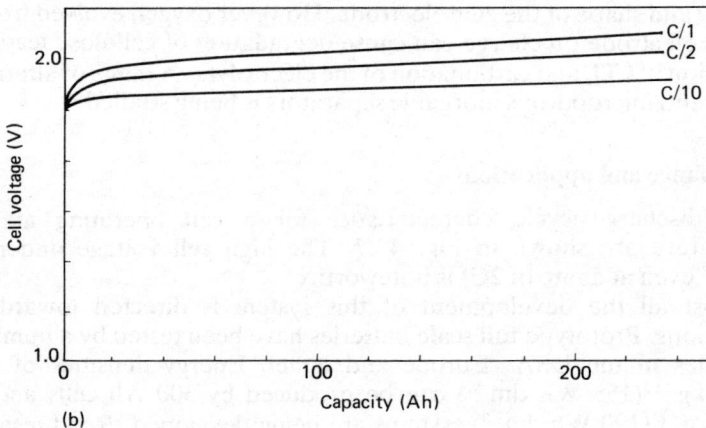

4.27   Discharge (a) and charge (b) characteristics of a typical 200 Ah zinc–nickel oxide cell as a function of rate.

species) to the zinc electrode by using a cellophane membrane separator. Later developments of this system have emphasised the fundamental role played by separator characteristics in producing a cell of acceptable reliability and cycle life. Apart from high cost and relatively poor cycle life, zinc–silver oxide cell performance deteriorates rapidly below 10°C. However at higher temperatures the cells can sustain high discharge currents without significant polarisation, and their high energy density makes them very suitable for aerospace and certain military applications.

The fully charged cell may be written as

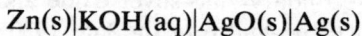

$$Zn(s)|KOH(aq)|AgO(s)|Ag(s)$$

The cell reaction takes place in two stages:

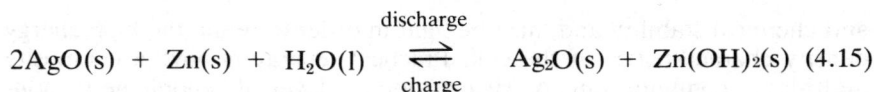

$$2AgO(s) + Zn(s) + H_2O(l) \underset{\text{charge}}{\overset{\text{discharge}}{\rightleftharpoons}} Ag_2O(s) + Zn(OH)_2(s) \quad (4.15)$$

and

$$Ag_2O(s) + Zn(s) + H_2O(l) \underset{\text{charge}}{\overset{\text{discharge}}{\rightleftharpoons}} 2Ag(s) + Zn(OH)_2(s) \quad (4.16)$$

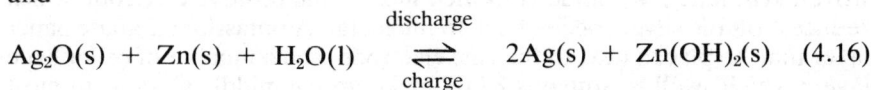

with associated e.m.f. values of 1.85 V and 1.59 V at 25 °C respectively. Both zinc and silver soluble species are also involved in the cell process.

Cell construction

While there are a number of methods used for manufacturing the positive electrodes, the two most important processes are (i) the sintering of silver powders and (ii) slurry pasting. The former procedure produces electrodes with superior mechanical properties. The silver mass which is formed by sintering silver powder at temperatures between 400°C and 700°C, is supported on silver or silver plated copper grids. This method allows a continuous manufacturing process and a final assembly with the positive electrode in a fully discharged condition – which greatly improves the storage characteristics of the system. Organic resins and pore-forming materials may be added to the silver and continuous sheets formed by a rolling process. This method is used when thin sintered plates are required. Non-sintering methods are based on electroforming plates which have been prepared by spreading a silver oxide/water slurry onto suitable support grids. In some processes, thermal decomposition at 400°C precedes electro-forming. Again, this method of manufacture may be performed on a continuous basis.

Many different fabrication procedures are used for the zinc electrode. One group of methods starts with zinc oxide and various additives and follows with electrochemical reduction in dilute aqueous KOH. In another, zinc is electrodeposited from a cyanide bath. Others use mixtures of zinc, zinc oxide and organic binding agents. Generally the electrodes are formed on open-mesh grids. In all cases the aim is to produce electrodes of high porosity and controlled thickness. Additives include surface active agents to minimise dendritic growth and mercuric ions to increase the hydrogen overvoltage of the zinc and so reduce corrosion.

The separator assembly is the most critical component of zinc–silver oxide secondary cells. In addition to its normal function of preventing contact and short circuit between electrodes of different polarity, a separator in this system must also

(i) prevent silver migration to the negative electrode,
(ii) control zincate migration,
(iii) have the swelling properties required to establish favourable zinc morphology and distribution on charge, and
(iv) preserve the integrity of the zinc electrode.

In addition the separator must have a low electrical resistance, good thermal and chemical stability and must be light in order to retain the high energy density characteristics of the cell. Practical separators have a composite multi-layer configuration. A 'silver-stopping' layer of cellophane or non-woven synthetic polyamide is located next to the positive electrode which reduces soluble silver species back to the metal. A potassium titanate paper layer may be placed next to the zinc electrode, and a number of cellophane layers which swell in aqueous KOH make up the middle section. In most cells the separators are fabricated as envelopes or 'sacks' which completely enclose the zinc electrodes.

Commercial cells are generally rectangular (prismatic) in shape and the case is usually plastic, with a rugged construction to withstand the mechanical stress to which this system is often subjected. The cells are sealed, but normally have a safety filler valve assembly. (Small quantities of oxygen evolved at the silver electrode during recharge may recombine with excess zinc on the negative electrode: however the membrane separator assembly slows down oxygen diffusion.) The quantity of free electrolyte is very small: the majority is contained within the electrode pores or separator.

Freshly assembled cells are subjected to a series of formation cycles in order to activate the system. Cells are often sold in a charged but dry state, in which case the formation process is performed before the final cell assembly. Dry charged batteries can be stored indefinitely. Cells containing electrolyte should be stored in the completely discharged state since dissolution of silver oxide is then avoided.

Performance and application

The energy density of practical zinc–silver oxide cells is some 5–6 times higher than that of their nickel–cadmium equivalents. In Fig. 4.28(a) typical low and high rate discharge curves are shown. At low rates two plateaus appear, at about 1.7 V and 1.5 V, corresponding to the reactions given in equations 4.12 and 4.13 respectively. The change from Ag(II) to Ag(I) controlled processes is dependent on the discharge rate and the upper plateau disappears completely at high rates. Reduction in cell capacity at high rates of discharge is not very significant. However sustained high rates can result in temperatures which may damage organic separator materials. In Fig. 4.28(b) a typical recharge curve at C/10 is shown (much faster charging is possible). The cell can be considered to be fully charged when its voltage reaches 2.0 V – excessive overcharge should be avoided because of the increased probability of zinc dendrite formation and possible damage to the separator by reaction with oxygen. The main drawbacks of the system are its high cost combined with a poor working and cycle life.

Zinc–silver oxide secondary cells with capacities of 0.5 to 100 Ah are manufactured for use in space satellites, military aircraft, submarines and for supplying power to portable military equipment. In space applications, the batteries are used to augment the power from solar cells during periods of high demand – e.g. during radio transmission or when the sun is eclipsed. At other times the batteries are trickle or float charged by the solar cells. The battery for the 'Ranger' lunar photography spacecraft, which consisted of

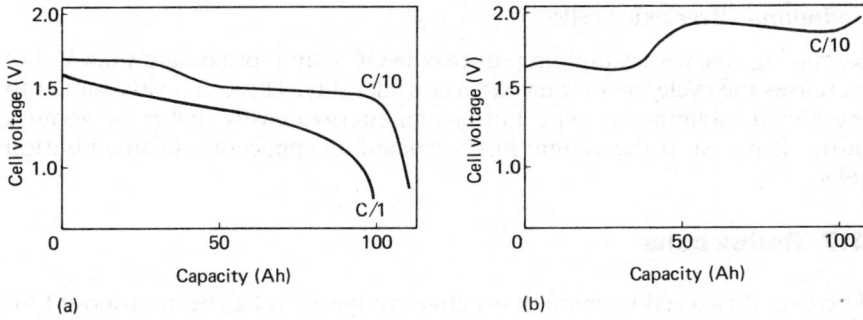

4.28 Discharge (a) and charge (b) characteristics of a typical 100 Ah zinc–silver oxide cell.

fourteen sealed 45 Ah cells is shown in Fig. 4.29. This spacecraft impacted on the moon in 1965. In military aircraft the use of zinc–silver oxide batteries as emergency power supplies or to provide additional power for certain manoeuvres requiring rapid actuation of flight control surfaces, may be justified because of their high energy density. 'One-shot' reserve zinc–silver

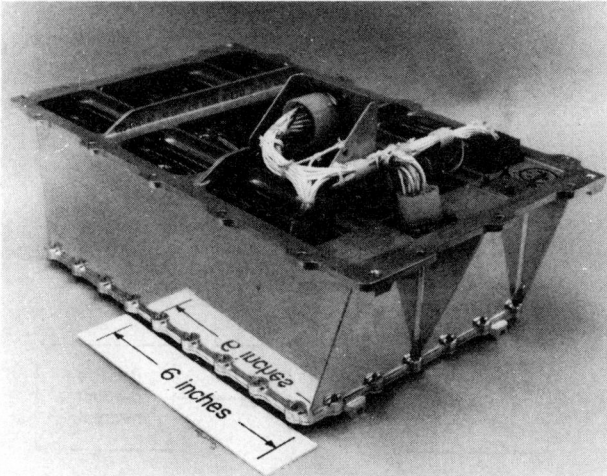

4.29 Battery for 'Ranger' lunar photography spacecraft, comprised of 14 sealed 45 Ah zinc–silver oxide cells, which impacted on the moon in 1965. (By courtesy of Jet Propulsion Laboratory)

oxide batteries are also used in military aircraft. A few large submarine batteries have also been built using the zinc–silver oxide system as an alternative to lead–acid cells. Heat dissipation problems require that particular attention be given to the way in which the cells are stacked. Submarine batteries of this type are usually stored under oil.

Cadmium–silver oxide cells

Replacing zinc with cadmium reduces the OCV by approximately 0.4 V, but increases the cycle life of the system considerably. This cell is very similar to the nickel–cadmium system, but has an energy density higher by about a third. The cost of the system has restricted its application to small button cells.

4.7 Redox cells

The use of two redox couples as active components has been proposed for cells designed for load levelling and bulk energy storage. Such systems have inherently low energy densities and are not likely to be suitable for other applications. Two types of cell system are being considered.

(i) Flow cells

This form of cell is shown schematically in Fig. 4.30. The anolyte and catholyte are different redox solutions which flow or are pumped past inert electrodes. The cell is constructed of two compartments separated by an

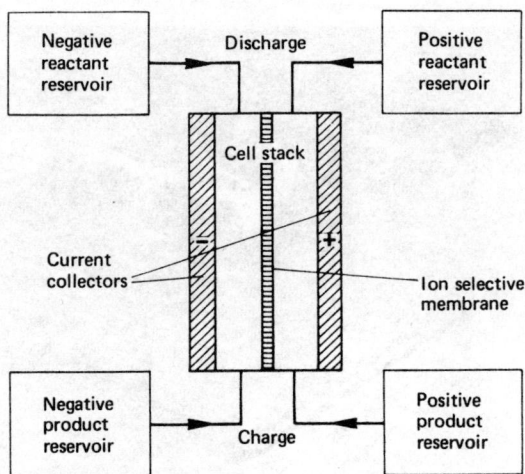

4.30   Schematic cross-section of a redox flow cell.

anion-selective semipermeable membrane. The spent solutions are returned to storage tanks and the whole process is reversed during charge. The general cell reaction is thus

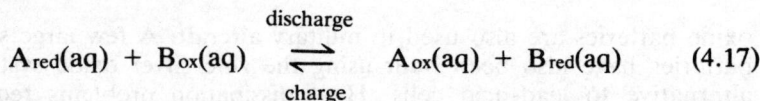

$$A_{red}(aq) + B_{ox}(aq) \underset{charge}{\overset{discharge}{\rightleftharpoons}} A_{ox}(aq) + B_{red}(aq) \qquad (4.17)$$

The system has a number of attractive features, especially the flexibility of capacity and power output: capacity can be increased simply by enlarging the size (or number) of the storage tanks; power output can be raised by increasing the flow rate or by bringing into line parallel cell sections (cf. fuel cells). Further advantages include low operating temperatures and the absence of discharge depth or cycle life limitations.

A screening of possible redox couples has led to the choice of Fe^{3+}/Fe^{2+} as a very suitable catholyte, while Cr^{2+}/Cr^{3+} and Ti^{3+}/TiO^{2+} have been suggested for the anolyte. The latter shows problems connected with hydrolysis effects unless very low pH is maintained, and in addition has a low exchange current density. Cr^{2+}/Cr^{3+} is also kinetically slow, especially in the charging direction. In cells tested so far, voltages and current densities have been low and considerable improvements are needed in electrode performance and in the development of anion-selective membranes with lower resistivity. Theoretical models of 10 and 100 MWh systems have been developed by NASA/ERDA in the USA.

(ii) Cells with insoluble redox couples

A system has been proposed which involves quinone/hydroquinone interconversions. A number of such couples are insoluble in acid solution and can therefore be used as active solid masses when made conductive by the addition of graphite. A system which has been studied experimentally is based on anthraquinone, Q, and tetrachloro-p-benzoquinone (chloranil), Q′ and their reduced or hydroforms. The overall cell reactions are then

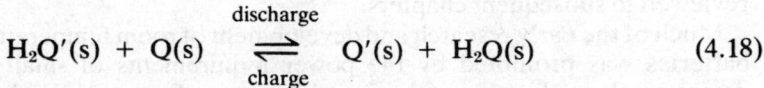

$$H_2Q'(s) + Q(s) \underset{charge}{\overset{discharge}{\rightleftharpoons}} Q'(s) + H_2Q(s) \tag{4.18}$$

No solution flow or ion-selective membrane is required and the volume of electrolyte required is low. The total amount of acid is invariant since only proton transfers are involved. The OCV of the cell is 0.6 V and the theoretical energy density is 67 Wh kg^{-1}.

5 Ambient lithium liquid electrolyte cells

5.1 Introduction

Rapid developments of semiconductor and display technologies and the exploitation of intermittent energy sources such as solar energy have made increasing demands on battery technology for systems with higher specific capacity, energy and power. Of all possible anode materials, lithium is perhaps the most attractive since it combines a favourable thermodynamic electrode potential with a very high specific capacity ($3.86\,\mathrm{Ah\,g^{-1}}$; $7.23\,\mathrm{Ah\,cm^{-3}}$). As a result of its electropositive nature, lithium rapidly reduces water, and cells with lithium anodes generally employ non-aqueous electrolytes. In this chapter, systems operating at ambient temperatures and based on salts dissolved in aprotic organic solvents or in liquid cathodes will be discussed; lithium cells with molten salt or solid electrolytes will be reviewed in subsequent chapters.

Much of the early research and development of room temperature lithium batteries was promoted by the power requirements of small electronic devices such as electric watches, cardiac pacemakers, and hearing aids and of military and aerospace equipment. More recent research has focused on the problems of expanding the field of application of lithium–organic and similar systems through the developments of high capacity rechargeable batteries with characteristics similar to those of the high temperature sodium–sulphur and lithium–iron sulphide systems. In Fig. 5.1 the specific energy and power ranges of lithium-based ambient batteries are compared with those of other primary and secondary systems. The superior values of specific energy are evident. Specific power is, however, limited, mainly because of the relatively poor conductivity of the electrolytes. On the other hand, their low freezing point allows a lower working temperature than that for conventional aqueous batteries and the low chemical reactivity of lithium with these electrolytes leads to a satisfactory shelf life.

Lithium systems may be classified according to the physical state of the positive electroactive material:

(i) **solid cathode reagents:** compounds with a negligibly small solubility in the electrolyte, e.g. CuO, FeS,

(ii) **soluble cathode reagents:** the only important example is sulphur dioxide, and

(iii) **liquid cathode reagents:** the active species is in liquid form at the cell operating temperature, e.g. thionyl chloride, $SOCl_2$ and sulphuryl chloride, SO_2Cl_2.

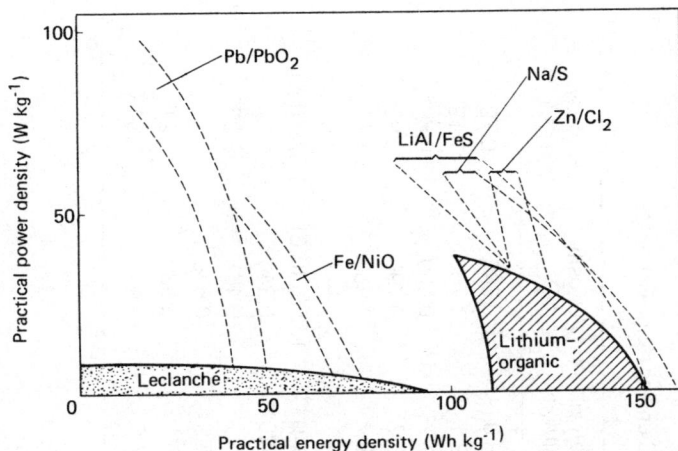

5.1 Power density – energy density curves for practical battery systems.

Electrolytes for (i) and (ii) are lithium salts dissolved in aprotic solvents. In (iii) lithium salts are added to the liquid electroactive molecules as 'supporting electrolytes' to carry the current through the cell. Before describing a number of primary and secondary cells of technical and commercial importance, the general properties of the electrolytes, the lithium anode and the more common cathode materials will be discussed.

5.2 Electrolytes

Selection of the most suitable solute-solvent combination for a battery electrolyte involves consideration of the conductance of the resulting solution, its chemical and electrochemical stability and its compatibility with the electrode materials. In the case of lithium batteries, lithium inorganic salts dissolved in aprotic organic solvents have proved to be the most suitable electrolytes. The most commonly used solvents include cyclic esters (ethylene carbonate, propylene carbonate, γ–butyrolactone), linear esters, amides and sulphoxides: physical properties of these materials are listed in Table 5.1. **A mixed solvent** is sometimes preferred since the properties of the electrolyte solution (conductance, viscosity, etc.) and its reactivity towards lithium can often be 'tailored' to give optimum performance. This is particularly important in the case of secondary systems (see Section 5.5) whose success depends largely on the suitability of the electrolyte for recharging the lithium electrode, and for subsequent charge retention. Choice of solute is limited mainly by solubility, which must be high so that the resulting solution will have sufficient conductivity and hence the cell will have a relatively low internal resistance. The most widely used salts are $LiClO_4$, $LiAlCl_4$, $LiBF_4$ and $LiAsF_6$ – i.e. either simple salts or combinations of a lithium halide with a Lewis acid.

Table 5.1

Properties of some solvents commonly used in lithium–organic cells (at 25°C unless otherwise state)

Solvent	Abbreviation	Molec. mass	m.p. °C	b.p. °C	Relative Permittivity	Viscosity cP†	Density g cm^{-3}
Acetonitrile	AN	41.05	−45.7	81.6	38.0	0.345	0.79
γ-butyrolactone	BL	86.09	−42	206	39.1	1.750	1.13
1,2-dimethoxyethane	DME	90.12	−58	83	7.2	0.450	0.86
N,N-dimethylformamide	DMF	73.10	−61.0	149	36.7	0.796	0.95
Dimethylsulphoxide	DMSO	78.13	18.5	189.0	46.7	1.960	1.10
Dioxolane	DIOX	74.08	−95.0	78.0	—	—	1.06
Ethylene carbonate	EC	88.06	36.4	248	89.6 (40°C)	1.850 (40°C)	1.32 (40°C)
Methyl formate	MF	60.05	−99.0	31.5	8.5 (20°C)	0.340	0.97
2-methyltetrahydrofuran	MeTHF	86.12	−137.0	80	—	0.461	0.880
Nitromethane	NM	61.04	−28.5	101.0	35.9	0.620	1.14
Propylene carbonate	PC	102.09	−48.8	242	66.1	2.530	1.21
Tetrahydrofuran	THF	72.12	−108.0	65.0	7.4	0.457	0.848

† 1 cP = 0.001 kg m^{-1} s^{-1}

The conductance of lithium salt solutions in aprotic solvents generally shows a maximum as the concentration of electrolyte is increased, as illustrated in Fig. 5.2. Such maxima can be interpreted on the basis of the opposing influence of an increasing number of charge carriers on the one hand, and (a) increasing viscosity and (b) increasing ion association with the

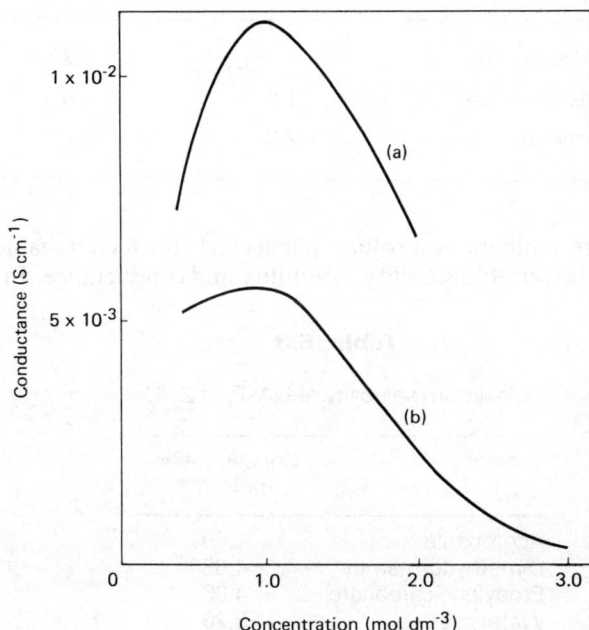

5.2 Conductance of $LiAsF_6$ in γ-butyrolactone (a), and propylene carbonate (b) as a function of concentration. (By permission of the Electrochemical Society)

formation of non-conducting ion pairs, on the other. It has recently been shown that the conductance can be increased by the addition of crown ethers, such as 12–crown–4:

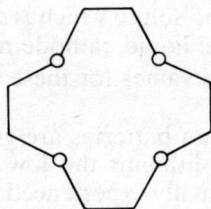

This can be attributed to a decrease in ion-association due to shielding of the charge on the lithium ions. The conductance of a number of $LiClO_4$ solutions are given in Table 5.2.

Table 5.2

Highest values of conductance of some $LiClO_4$ solutions at 25°C

Solvent	Concentration $mol\,dm^{-3}$	Conductance $S\,m^{-1}$
Dimethylformamide	1.16	2.22
Dimethylsulphoxide	1.5	1.0
Methyl formate	2.8	3.20

$LiAsF_6$ is pre-eminent as a solute, particularly for rechargeable systems, because of its favourable stability, solubility and conductance. In Table 5.3

Table 5.3

Maximum solubility of $LiAsF_6$ at 25°C

Solvent	Concentration $mol\,dm^{-3}$
Acetonitrile	1.60
Dimethylformamide	4.68
Propylene carbonate	4.66
Water	3.28

the solubility of this salt in a number of solvents is reported. In methyl formate (MF), $LiAsF_6$ dissolves to give one of the most conductive organic electrolyte solutions known: at 25°C a solution of $2\,mol\,dm^{-3}$ has a conductance of $4.5\,S\,m^{-1}$. While this combination can be considered as a suitable electrolyte for 'high rate' cells, it suffers from some decomposition with gassing in the presence of lithium unless stabilised with materials such as $(CH_3)_4NBF_4$. $LiAlCl_4$ is the solute which is most widely used to increase the conductance of inorganic liquid cathode materials such as $SOCl_2$ and SO_2Cl_2. Typical conductance values for these solutions are given in Table 5.4.

The solvents used in lithium batteries are generally thermodynamically unstable in the presence of lithium: the low lithium corrosion rate and consequent good shelf life actually experienced with sealed cells is due to the formation of a protective film on the surface of the metal. The practical stability of the electrolyte solutions in the presence of lithium depends on their purity and in particular, on low water content. A number of procedures have been proposed for the purification and drying of solvents, including operations such as pre-electrolysis with platinum electrodes. In most cases it

Table 5.4

Conductivity of some LiAlCl$_4$ solutions in organic solvents at 25°C

Solvent	Concentration mol dm^{-3}	Conductance S m^{-1}
SOCl$_2$	2.0	2.0
SO$_2$Cl$_2$	2.0	1.0

has proved sufficient to place the solvent in contact with molecular sieves for a few days before fractional distillation.

The electrochemical stability of a solution can be defined as the voltage range within which it does not undergo detectable electrolytic decomposition, for a particular pair of electrodes. In the case of a practical primary cell such a range must be wider than the OCV. For secondary systems, the value of the voltage applied to the cell during recharge must also be considered. The composition and history of the electrodes have an important influence on the decomposition voltage. This illustrates the importance of kinetic over thermodynamic factors. It is often difficult to establish whether a decomposition current is due to reactions involving solvent or solute, but in practice a marked influence of the nature of the solute is often observed. The oxidation voltages with respect to lithium of some related electrolytic solutions used in lithium batteries are given in Table 5.5. These values were determined using linear scan voltammetry with platinum electrodes.

Table 5.5

Oxidation potentials of some inorganic salt/ether solvent systems

Solvent	Salt	Concentration mol dm^{-3}	Oxidation Potential vs Li$^+$/Li
Dioxolane	LiClO$_4$	2.0	3.4
	LiPF$_6$	1.0	3.50
	LiAsF$_6$	1.5	3.60
Dioxolane + 0.1 mol dm^{-3}			
pyridine	LiAsF$_6$	1.5	3.84
2–Methyltetrahydrofuran	LiAsF$_6$	1.0	4.32
4–Methyltretrahydrofuran	LiAsF$_6$	1.0	4.32

Room temperature; platinum substrate; current density 1 mA cm^{-2}. (H.H. Horowitz et al., Fall Meeting of the Electrochemical Society, Hollywood, Florida, 1980. Extended Volume 80–2 (1980), p. 177) Reprinted by permission of the publisher, The Electrochemical Society, Inc.)

5.3 The lithium anode in primary cells

The properties of lithium are summarised in Table 5.6. The metal is both ductile and malleable. It is harder than the other alkali metals, but softer than lead. It shows a tendency to stick to various materials under moderate pressure. In humid air the freshly cut shiny surface of lithium rapidly turns greyish-white to grey or black, due to a reaction with oxygen and nitrogen.

Table 5.6

Physical properties of lithium

Atomic weight	6.94
Melting point	180.5°C
Boiling point	1347°C
Density	$0.534\,g\,cm^{-3}$ at 20°C
Heat of fusion	$3.001\,kJ\,mol^{-1}$
Heat of vapourisation	$147.1\,kJ\,mol^{-1}$
Resistivity	$9.446 \times 10^{-6}\,\Omega\,cm$ at 20°C

However if the humidity is maintained at a low enough level ($<2\%$ at 20°C) there is no apparent tarnishing for periods of hours. 'Dry rooms' can be constructed for the assembly of lithium batteries. The handling of lithium is comparatively safe in comparison with the other alkali metals: for example, it will not spontaneously ignite, unless in finely divided form and in contact with water.

Film formation on lithium

The reaction of lithium with the electrolyte to form a surface film significantly modifies its behaviour. On the one hand the film confers chemical stability and useful shelf-life on the system. On the other, it is responsible for greatly depressed exchange currents and the consequent phenomenon of voltage delay, as discussed in Chapter 3 in connection with magnesium aqueous batteries. It is convenient to discuss separately film formation with (a) insoluble and (b) liquid and soluble cathode systems.

(a) Cells with insoluble cathodic reagents

In these cells, provided that the solubility of the cathode material is very low, the solvent itself is principally responsible for film formation. The most detailed studies have been performed on PC-based electrolytes where the thermodynamically favoured reaction is:

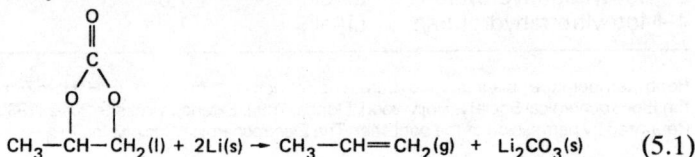

$$CH_3-CH-CH_2\,(l) + 2Li(s) \rightarrow CH_3-CH{=}CH_2\,(g) + Li_2CO_3(s) \qquad (5.1)$$

Propene evolution has been observed on lithium amalgams and also at platinum surfaces connected to lithium electrodes. With solid lithium itself, however, a lithium carbonate film is immediately formed which passivates the metal surface. The level of trace impurities, such as water, does not affect the film formation significantly. Passive films have also been observed on lithium in solvents such as BL, DME, AN, MF, THF, etc.

On drawing current from a passivated lithium anode, polarisation is at first severe, but the voltage recovers fairly rapidly (Fig. 5.3). Initially, the charge transfer process at the anode is mainly lithium dissolution through imperfections in the film. This dissolution progressively disrupts and removes the film, thus allowing the discharge voltage to rise rapidly to its

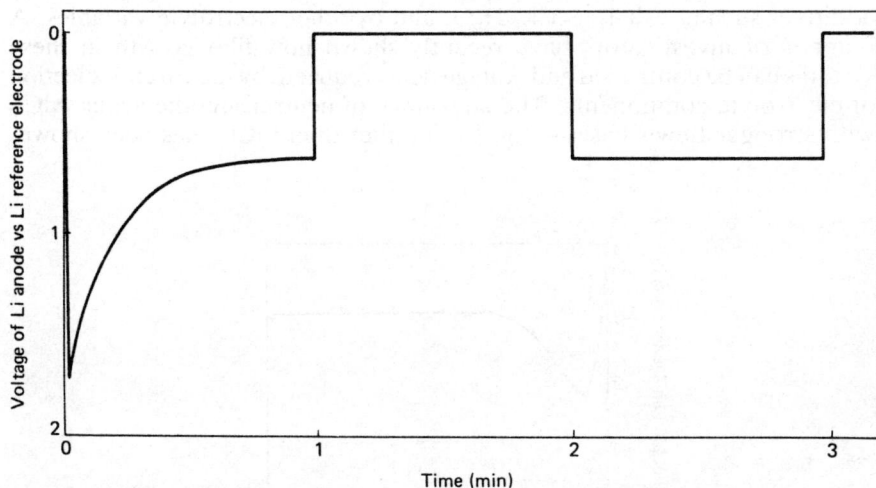

5.3 Voltage recovery of a lithium anode at −20 °C in 1 mol dm^{-3} LiClO$_4$ in PC versus a lithium reference electrode. Current density: 10 mA cm^{-2}.

steady value. If the cell is now returned to open circuit, and a second current pulse is subsequently withdrawn, no voltage delay is observed if the period on open circuit is too brief for reformation of the film. In commercial lithium batteries of this type, recovery from voltage delay is very rapid under normal working conditions, but is slower at low temperatures.

(b) Cells with liquid or soluble cathodic reagents

The formation of passivating films on lithium in contact with liquid or soluble cathodic reagents is a pre-requisite for the construction of a practical cell. The film acts in the same way as a separator, preventing further direct chemical reaction of lithium and the cathodic reagent. However film formation involving the action of SO$_2$, SOCl$_2$, etc. on lithium is considerably more complex, and may produce much more severe voltage delay characteristics than in the case of insoluble cathodes described above.

When SO_2 dissolved in AN is brought into contact with lithium, a layer of lithium dithionite is formed, following the same reaction scheme as the normal cell reaction:

$$2Li(s) + 2SO_2(diss.) \rightarrow Li_2S_2O_4(s) \tag{5.2}$$

If the solvent is a PC/AN mixture, the film constitution is complex and contains Li_2CO_3 in addition to $Li_2S_2O_4$. This latter film is probably more coherent or compact since the voltage delay is greater than for the pure dithionite film. With $SOCl_2$ and SO_2Cl_2 containing $LiAlCl_4$ a particularly severe voltage delay is observed, especially after prolonged storage, as shown in Fig. 5.4. The film consists primarily of LiCl crystals whose growth and morphology follow a complex pattern and are affected substantially by additives such as $AlCl_3$, S_2Cl_2, H_2O, and by other electrolyte variables. A number of investigators have recently shown how film growth in these systems can be controlled and voltage delay reduced, by the careful selection of electrolyte components. The advantage of neutralising the acidic $AlCl_3$ with stronger Lewis bases – e.g. Li_2O rather than LiCl – has been shown.

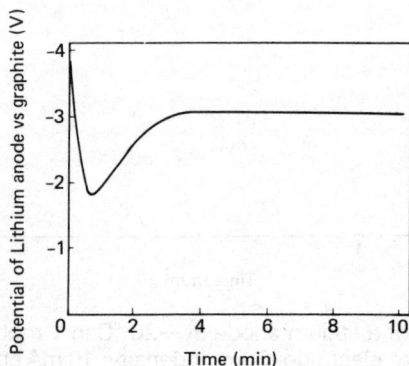

5.4 Initial discharge behaviour of a Li(s)|$LiAlCl_4$,SO_2Cl_2(l)|C(s) cell after prolonged storage at room temperature. Current density at lithium electrode: $5\,mA\,cm^{-2}$.

The effect of the change of the Lewis acid-base properties of the electrolyte on the voltage delay are shown in Fig. 5.5. The influence of alloying the lithium anode has also been studied.

In practical terms, the twin objectives of protecting the lithium from corrosion while avoiding unacceptable levels of voltage delay can be considered to have been met. However the detailed mechanisms of film formation and disruption are still matters of some controversy. In particular, the interaction of thin films formed rapidly on lithium surfaces exposed to the atmosphere with the thicker films formed by subsequent reaction with the cathodic reagent, is not well understood.

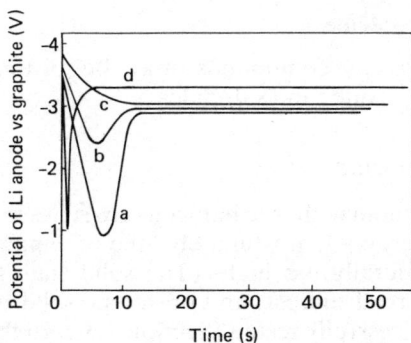

5.5 Effect of electrolyte composition on discharge behaviour of Li(s)|(electrolyte, 1 mol dm^{-3}), SOCl$_2$(l)|C(s).
   (a) AlCl$_3$.LiCl; (b) AlCl$_3$.Li$_2$S; (c) AlCl$_3$.Li$_2$O; (d) AlCl$_3$.LiF. Current density at lithium electrode: 6.4 mA cm^{-2}. Storage: 15 days at 25°C or 7 days at 70°C. (By permission of the International Society for Electrochemistry)

5.4 Cathode materials and lithium primary cells

Lithium primary cells can be divided into three categories depending on the type of cathode used. In the first group, the cathodic reagent is a solid material, almost completely insoluble in the electrolyte; in the second it is completely dissolved in the electrolyte; and in the third it is a liquid, and in addition to being the species responsible for the uptake of electrons from the external circuit, it acts as solvent for the supporting electrolyte which carries the current through the cell. Most of the commercial cells currently on the market belong to the first group, although there are examples in the other two categories. The cells are manufactured in a number of forms of which the standard button cell (Fig. 5.9) is the most important. When cells with higher capacities are required, a spiral wound design, as shown schematically in Fig. 5.6, is preferred. Bobbin-type cells are used to achieve higher energy density.

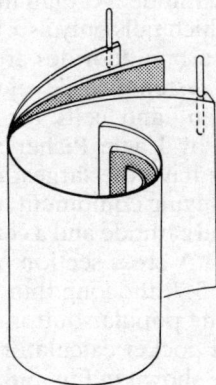

5.6 Spiral wound battery construction.

(a) Solid cathode systems

Four main groups of compounds may be distinguished: polycarbon fluorides, oxosalts, oxides and sulphides.

(i) *Polycarbon fluorides*

The coupling of lithium with free halogens gives rise to cells of exceptionally high specific energy. Unfortunately the chemical reactivity of the components is generally too high. (The solid state lithium–iodine cell is however a commercial success, and prototype solid state lithium–bromine cells have been successfully tested (Section 7.3).) In the case of fluorine the problem has been overcome by effectively immobilising the fluorine in a graphite host. Polycarbon fluorides, of general formula $(CF_x)_n$ can be obtained by direct fluorination of carbon black, or other carbon varieties at high temperatures. For fluorine compositions in the range $0.4 < x < 1.0$, such subtances have very high specific energies (e.g., 2600 Wh kg^{-1} for $x=1$). For $x<1$ the materials are electronic conductors. Lithium cells with polycarbon fluoride cathodes have OCV values in the range of 2.8–3.3 V, depending on the exact formulation of the cathode material.

A typical cell may be written as

$$Li(s)|LiBF_4,PC–DME|(CF_x)_n(s)$$

with associated cell reaction

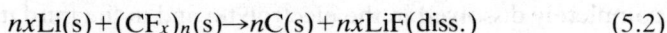

$$nxLi(s)+(CF_x)_n(s) \rightarrow nC(s)+nxLiF(diss.) \tag{5.2}$$

It is believed that the discharge mechanism involves the formation of an intermediate lithium 'intercalation' compound in which both lithium and fluorine are situated between the carbon layers of the graphitic structure.

The $(CF_x)_n$ cathode is stable in contact with organic electrolyte systems, and the cells have a satisfactory shelf life. In common with other lithium power sources, polycarbon fluoride-based cells suffer from voltage delay. However in these cells the effect may be more severe due to additional contribution from the cathode. At high fluorine contents, $(CF_x)_n$ has a high electronic resistance which falls only as x is reduced during cell discharge.

Cells based on polycarbon fluorides are manufactured commercially in a number of forms. The system was developed first by Matsushita Electrical Industrial Co. in Japan, and cells for military applications have been produced in the USA by Eagle Picher and Yardney Electric. The spiral wound cylindrical cells have the largest capacity and are used in portable radio transceivers, surveying equipment, etc. The so called 'inside-out' cells have a hollow cylindrical cathode and a central lithium anode, all enclosed in a light aluminium case. A cross-section of such a cell based on Matsushita design is shown in Fig. 5.7: the long thin form is an advantage in particular applications. Finally, the popular button cell is produced mainly for use in electronic watches and pocket calculators. Discharge curves of these cells under various loads are shown in Fig. 5.8. Table 5.7 gives the specifications of typical cells of each type.

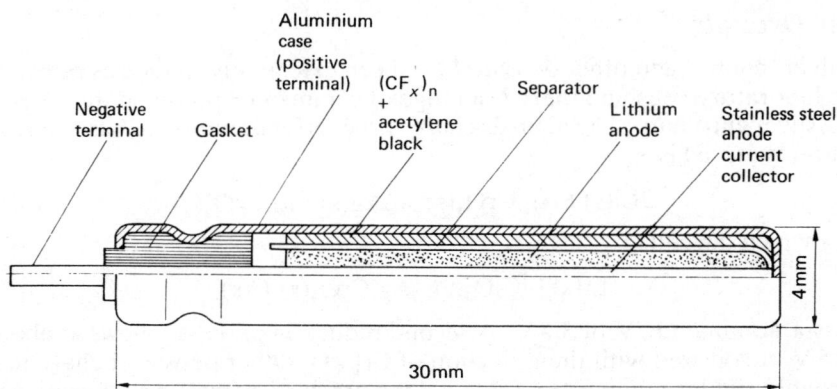

5.7   Lithium – $(CF_x)_n$ cylindrical cell (based on the design of Matsushita (BR435), by permission).

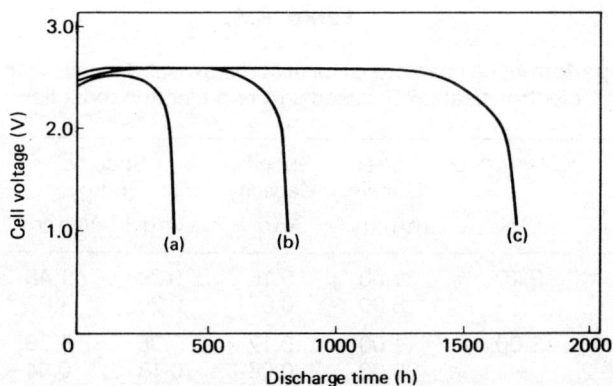

5.8   Discharge curves of lithium – $(CF_x)_n$ button cells under various loads at ambient temperature: (a) 5 kΩ; (b) 13 kΩ; (c) 30 kΩ. (By permission of Elsevier Sequoia, S.A.)

Table 5.7

Specifications of some commercial Li–$(CF_x)_n$ cells
(National Matsushita Electrical Industrial Co., Japan)

	Button	Inside-out	Spiral
OCV	3.0	3.0	3.0
Nominal capacity (mAh)	150	40	5000
Energy density (Wh kg^{-1})	140	140	320
Diameter (mm)	23	4.2	26
Height (mm)	2.5	35.9	50
Weight (g)	3.1	0.85	47.0

(ii) *Oxosalts*

Silver, copper and other oxosalts have been extensively studied as cathodes in laboratory cells and there is a range of commercial power sources based on silver chromate. The main discharge process for lithium cells based on the latter material is

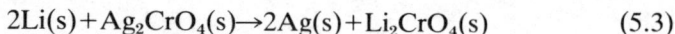

$$2Li(s) + Ag_2CrO_4(s) \rightarrow 2Ag(s) + Li_2CrO_4(s) \qquad (5.3)$$

for which the cell

$$Li(s)|LiClO_4, PC|Ag_2CrO_4(s), C(s)$$

has a nominal OCV of 3.5 V. A second reduction process follows at about 2.5 V, associated with the reduction of Cr(VI). Other oxosalts behave in a broadly similar manner. In Table 5.8 the practical energetic characteristics of a number of cathode materials coupled with lithium in $LiAsF_6$–BL cells are compared at two current densities.

Table 5.8

Practical performance characteristics of some oxosalts in cells with $LiAsF_6$/BL electrolyte at 25°C, based on a two-electron reduction.

Salt	OCV vs Li⁺/Li	Current Density	Specific Capacity	Specific Energy		Utilization to 1V cut-off
	V	mA cm⁻²	Ah g⁻¹	Wh g⁻¹	Wh cm⁻³	%
Ag_2WO_4	3.20	1.00	0.10	0.29	1.48	95
		8.00	0.07	0.20	1.02	75
Ag_2MoO_4	3.00	1.00	0.12	0.36	1.78	95
		8.00	0.06	0.13	0.64	50
$CuWO_4$	3.20	1.00	0.16	0.35	1.64	90
		5.00	0.16	0.28	1.35	60
		8.00	0.13	0.15	0.72	40
$CuMoO_4$	3.25	1.00	0.29	0.46	1.36	98
		5.00	0.21	0.23	0.68	90
		8.00	0.13	0.19	0.58	50

Silver chromate-based cells are manufactured in button and rectangular (prismatic) form in a number of sizes. The energy density of such complete systems is estimated as 200 Wh kg⁻¹ or 575 Wh dm⁻³, to a 2.5 V cut-off. A schematic cross-section of a typical button cell of this type is shown in Fig. 5.9 and Table 5.9 gives the dimensions and characteristics of the range of button cells produced by SAFT. An exploded view of a rectangular cardiac pacemaker cell is given in Fig. 5.10. The excellent reliability under continuous drain, and the low self-discharge (<1% per annum) characteristic of these cells have made them one of the commonest power supplies for pacemakers and other implanted devices such as neurological

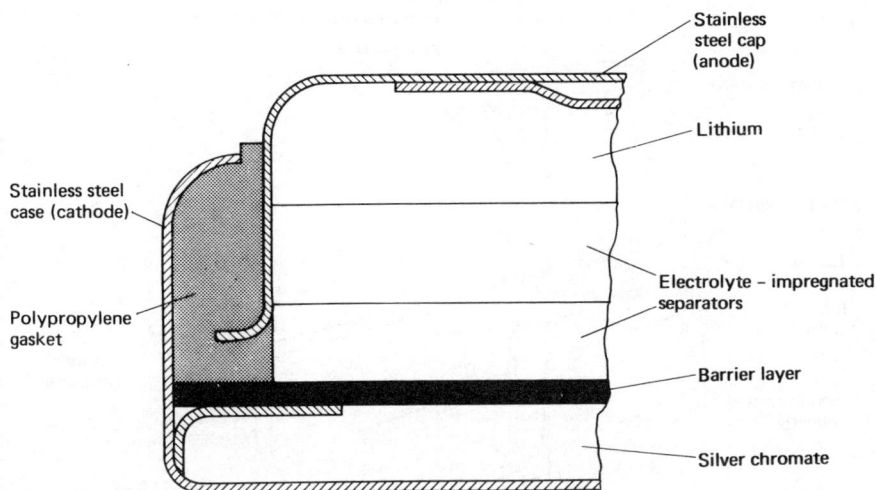

5.9 Cross-section of a typical lithium-silver chromate button cell. (Courtesy of SAFT Gipelec)

Table 5.9

Specifications of some commercial Li–Ag$_2$CrO$_4$ button cells (SAFT Gipelec)

OCV	3.5	3.5	3.5	3.5	3.5	3.5
Nominal capacity (mAh)	90	600	900	1310	1750	2300
Practical capacity at 2.7 V cut-off/mAh	86	540	810	1180	1570	2070
Diameter (mm)	11.4	21.0	27.3	35.5	35.5	35.5
Height (mm)	5.4	9.1	7.9	6.0	8.0	10.0
Weight (g)	1.7	8.9	12.5	18.0	22.5	29.0

pain-relieving systems, drug dispensers, etc. (The first cardiac pacemaker based on an Li/Ag$_2$CrO$_4$ cell was implanted in 1974.) Other uses include protection devices for electronic memories.

Typical accelerated and normal discharge curves for nominal 600 mAh button cells are shown in Fig. 5.11 and 5.12. As can be observed, some voltage delay is experienced at high current values. The characteristic two plateau form of the curve is used to give warning of the impending need for cell replacement. Generally the cells are formulated in such a way that anode limiting occurs when the second plateau has extended about a quarter of the length of the first, and a 2.5 V cut-off is taken in the determination of practical capacity.

Hermetic negative feedthrough
Positive terminal pin
Header assembly
Upper gasket
Webrill separator
Lithium anode
Internal connections
Anode current collector
Cathode container
Silver chromate cathode
Barrier separator
Collar
Insulating retainer frame
Stainless steel case

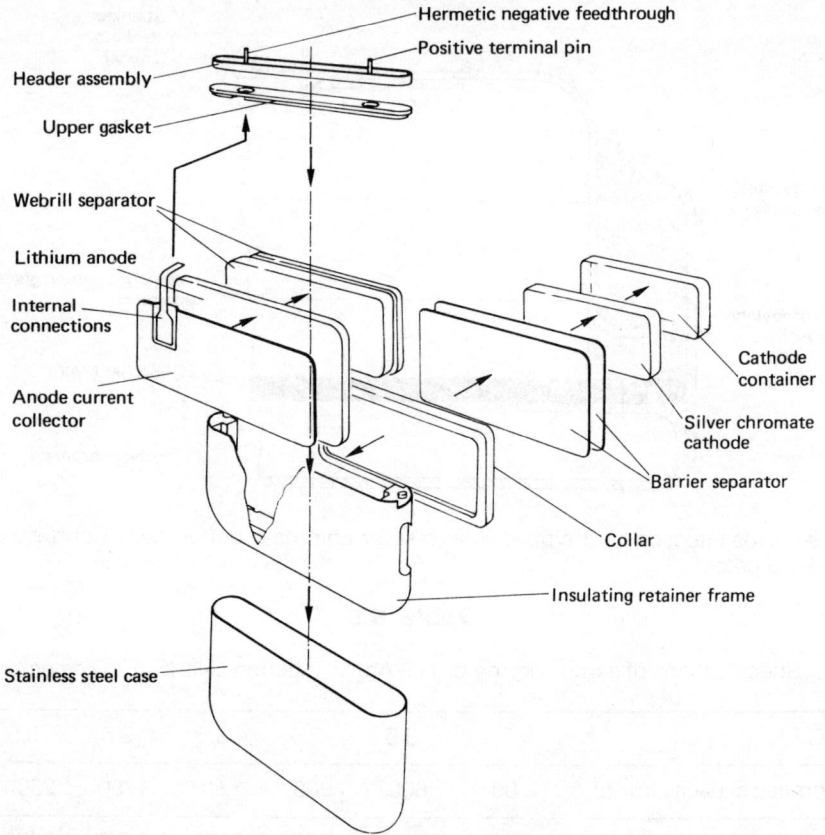

5.10 Exploded view of rectangular (prismatic) lithium–silver chromate pacemaker cell. (By courtesy of SAFT Gipelec)

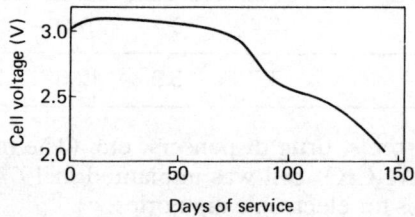

Cell voltage (V)

3.0

2.5

2.0

0   50   100   150

Days of service

5.11 Accelerated discharge curve for a lithium–silver chromate pacemaker cell of nominal capacity 600 mAh. (SAFT Gipelec Li210), under a load of 15 kΩ. (By courtesy of SAFT Gipelec)

A limited amount of swelling occurs during discharge, due to the positive volume change associated with the cell reaction. With some salts, such as Ag_2PO_4, this problem is more severe.

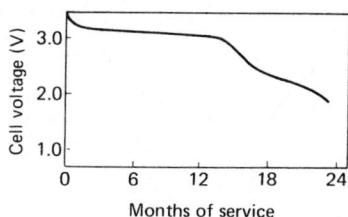

5.12 Discharge characteristics of a lithium–silver chromate pacemaker cell of nominal capacity 600 mAh (SAFT Gipelec Li210), under a load of 75 kΩ. (By courtesy of SAFT Gipelec)

(iii) *Oxides*

The discharge reaction for lithium–metal oxide cells is almost always more complex than the formal displacement process:

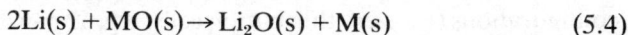

$$2Li(s) + MO(s) \rightarrow Li_2O(s) + M(s) \tag{5.4}$$

For example, Bi_2O_3 and PbO give rise to intermetallic compounds such as Li_3Bi and LiPb, while MoO_3 and V_2O_5 form ternary phases. In many cases (e.g. for the MnO_2 cathode) the discharge mechanisms are still not fully understood. In Table 5.10 theoretical capacities for a number of oxides are compared, assuming a straightforward displacement reaction. While a wide variety of metal oxides have been studied, cathodes of commercial cells are mainly confined to the oxides of manganese, copper, bismuth and lead.

Table 5.10

Theoretical capacities for a number of simple displacement reactions involving oxides

Active Cathode Material	Product	Electrons Involved	Capacity	
			Ah kg^{-1}	Ah dm^{-3}
Bi_2O_3	Bi	6	0.350	3.070
CuO	Cu	2	0.670	4.260
MnO_2	Mn_2O_3	1	0.310	1.550
Pb_3O_4	Pb	8	0.310	2.850

Two categories of cell may be distinguished: 'high voltage' and 'voltage compatible'. The latter term refers to the fact that lithium cells with discharge voltages of about 1.5 V can readily replace the more conventional miniature aqueous cells for which much electrical equipment has been designed.

The lithium–manganese dioxide high voltage cell introduced by Sanyo in 1975 is now manufactured by numerous companies, using specially prepared high density MnO_2. The OCV is in the range 3.0–3.5 V and typical discharge curves are shown in Fig. 5.13. Different sizes and forms of cell are produced

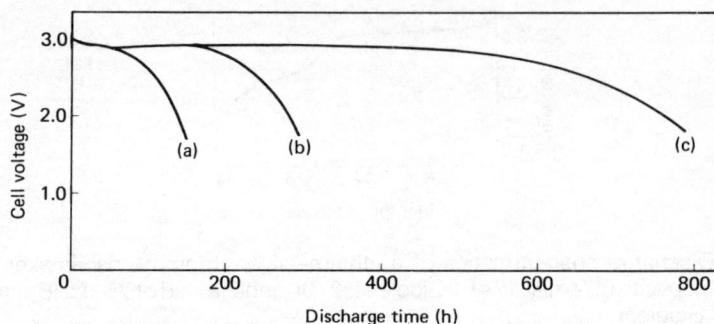

5.13 Discharge curves of lithium–manganese dioxide button cells (Varta CR 2025) under various loads at ambient temperature: (a) 2.7 kΩ; (b) 5.6 kΩ; (c) 15 kΩ. (Courtesy of Varta)

to suit particular applications. At present, cylindrical, button and 'thin coin' configurations (i.e. very thin button cells, e.g. 12 mm diameter $\times$ 2 mm high) are available with practical energy densities claimed to be in the range 500 to 700 Wh dm^{-3}. The discharge characteristics and low self-discharge make this a very suitable system for powering electric watches with LCD display. An important requirement for such a battery is the ability to provide current pulses to illuminate a miniature tungsten bulb (to enable the time to be read at night) while maintaining an acceptable voltage. Fig. 5.14 shows the behaviour of a SAFT LM 2020 cell under such a load. Another high voltage cell under development makes use of V$_2$O$_5$ as cathode material.

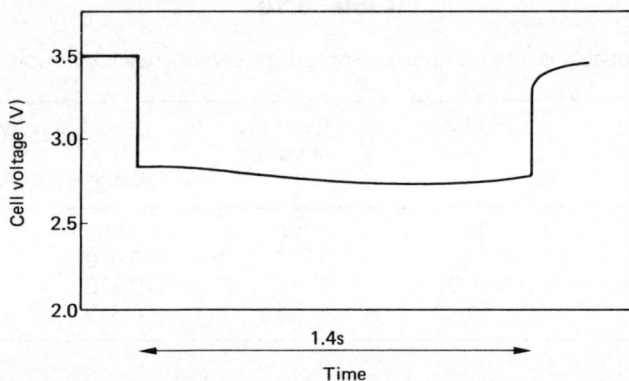

5.14 Response to high current discharge pulse of a typical lithium–manganese dioxide watch battery (LM 2020). Current: 6 mA. (By courtesy of SAFT Gipelec)

The lithium–copper oxide cell is voltage compatible (OCV $\approx$ 1.5 V) and has the highest specific energy of all solid cathode lithium-based cells. Practical values for complete cells of 750 Wh dm^{-3} (300 Wh kg^{-1}) are comparable with those of the lithium–sulphur dioxide system. The discharge curve shows a single step which may be attributed to the simple displacement reaction:

$$2Li(s) + CuO(s) \rightarrow Li_2O(s) + Cu(s) \qquad (5.5)$$

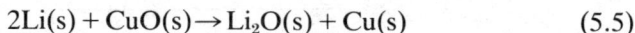

It has recently been shown that cathode utilisation and shelf-life are improved by incorporation of about 1 m/o of Li_2CO_3 or LiOH in the CuO, followed by heat treatment. ('m/o' is used to represent the molar percentage of a component in a mixture.) The electrolyte used varies from manufacturer to manufacturer: $LiClO_4$ in dioxolane has proved to be a very satisfactory system. Once again, self-discharge is very low. In Fig. 5.15 discharge curves after accelerated shelf testing at 70°C are shown. The cells behave well over

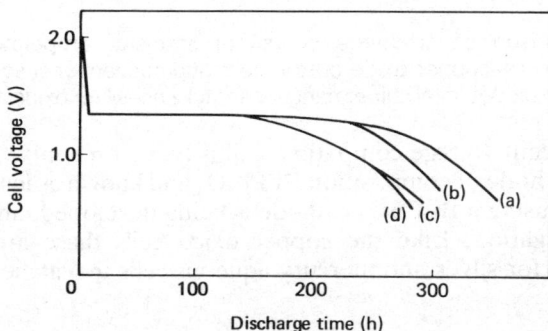

5.15 Discharge curves of lithium–copper oxide button cells (LC 01) after accelerated shelf testing at 70 °C. (a) fresh cell; (b) after 6 months at 70 °C; (c) after 12 months at 70 °C; (d) after 18 months at 70 °C. (By courtesy of SAFT Gipelec)

a wide temperature range at moderate current drain. In Fig. 5.16 the capacity of this type of cell is compared with an aqueous alkaline zinc cell of similar dimensions, under various conditions. The system has also been satisfactorily tested in resin encapsulated cells at temperatures as high as 150°C. The cells are manufactured in cylindrical form with theoretical

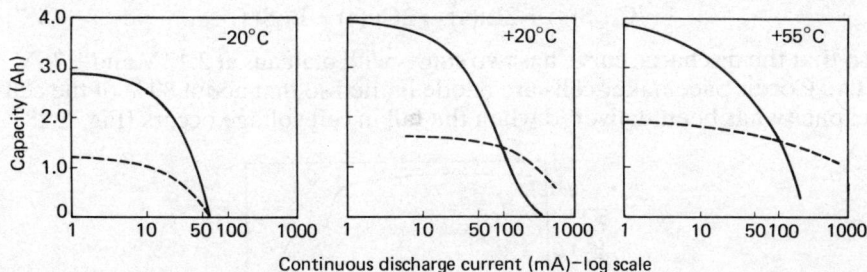

5.16 Capacity of lithium–copper oxide button cells (LC 01) as a function of discharge current (solid lines). Dashed lines indicate the characteristics of alkaline manganese cells of similar dimensions. (By courtesy of SAFT Gipelec)

capacities in the range 500–3900 mAh. A comparison of practical discharge capacities of lithium–copper oxide button cells and a number of aqueous alternatives is given in Fig. 5.17.

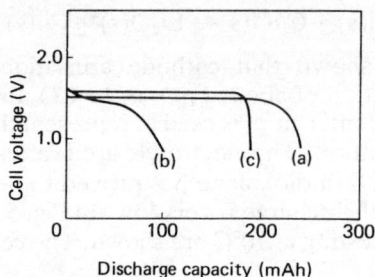

5.17 Comparison of discharge curves at ambient temperature of voltage compatible lithium–copper oxide button cells and conventional aqueous cells: (a) lithium–copper oxide; (b) alkaline manganese; (c) zinc–silver oxide. Load: 75 kΩ.

A more recent voltage compatible cell is based on a mixed bismuth and lead oxide cathode of composition $Bi_2Pb_2O_5$ and known as lead bismuthate. Another cell using a Bi_2CuO_4 cathode is being developed, and TiO_2 is also under investigation. Like the copper oxide cells these are intended as replacements for silver and mercury aqueous cells in watches, calculators, etc.

(iv) *Sulphides*

Metal sulphides have the advantage over the corresponding oxides that most of them are good electronic conductors and hence sulphide-based cathodes do not usually require the addition of carbon.

Batteries based on cupric sulphide cells (three in series) have been developed and used with cardiac pacemakers since 1976. Reduction of CuS takes place in two stages:

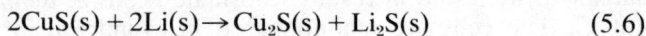

$$2CuS(s) + 2Li(s) \rightarrow Cu_2S(s) + Li_2S(s) \qquad (5.6)$$

and

$$Cu_2S(s) + 2Li(s) \rightarrow 2Cu(s) + Li_2S(s) \qquad (5.7)$$

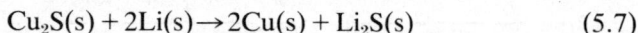

so that the discharge curve has two stages with plateaus at 2.12 V and 1.75 V. The Cordis pacemaker cells are anode limited so that about 80% of the cell capacity has been delivered when the fall in cell voltage occurs (Fig. 5.18).

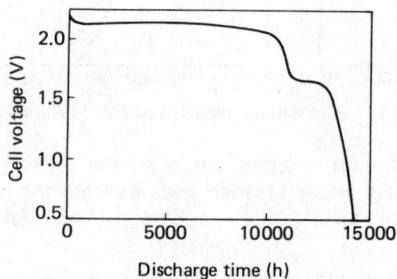

5.18 Discharge curve of a lithium–cupric sulphide pacemaker cell at 37 °C under a load of 12.3 kΩ. (By permission of the Electrochemical Society)

As in the case of the similar lithium–silver chromate cell described above, this voltage drop is used to indicate when battery replacement is required. The electrolyte in this cell is $LiClO_4$ dissolved in a mixed solvent, of which the major constituent is dioxolane. Self-discharge is negligible. An increase in volumetric capacity of 80% in comparison with aqueous mercury batteries is claimed together with an associated reduction in pacemaker weight.

Cells based on cuprous sulphide have also been developed: the continuous discharge characteristics of a Ray-O-Vac thin coin cell are shown in Fig. 5.19. Other promising systems include cells based on iron sulphide cathodes. The best performance has been produced with non-stoichiometric FeS_x, with $x \approx 1.1$.

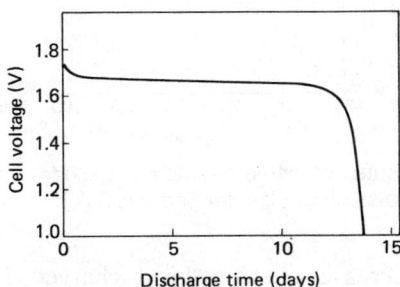

5.19 Discharge curve of a lithium–cuprous sulphide thin coin cell under a load of 12.5 kΩ. (By courtesy of INCO)

(b) Soluble cathode systems

The principle of using a 'soluble depolariser' is well established in aqueous cells. In the Grove cell of 1838 a zinc anode was coupled with nitric acid as cathodic reagent, using a platinum cathodic current collector. This system was later developed into the commercially successful Grove–Bunsen cell where a carbon current collector was substituted. In the Poggendorff or 'bichromate cell' of the same period, the catholyte was $Na_2Cr_2O_7$ in acid solution. The common feature of all successful cells of this type is a mechanism which prevents sustained attack on the anode by the cathodic reagent. In the case of lithium–organic cells, this mechanism derives from the formation of a passive layer on the metal, as described in Section 5.3 above.

All the currently available commercial lithium cells using a soluble cathodic reagent are based on sulphur dioxide. These advanced cells which have outstanding performances, but also certain drawbacks, have until recently been limited to military uses. Now they are finding increasing service for heavy duty applications in a wider consumer market.

The cell may be represented as:

$$Li(s) \; SO_2, LiBr, AN \; C(s),$$

with an overall reaction:

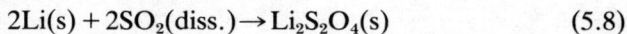

$$2Li(s) + 2SO_2(diss.) \rightarrow Li_2S_2O_4(s) \qquad (5.8)$$

An OCV of just over 3.0 V is observed. A typical electrolyte is a solution of LiBr in AN which has a conductance of 5 S m^{-1} at room temperature, falling to 2 S m^{-1} at $-50°$C. This rather small reduction leads to excellent low temperature discharge characteristics. The concentration of dissolved sulphur dioxide in an undischarged cell at room temperature gives rise to an internal pressure of about 300 kPa (3 atm). According to equation 5.8, the

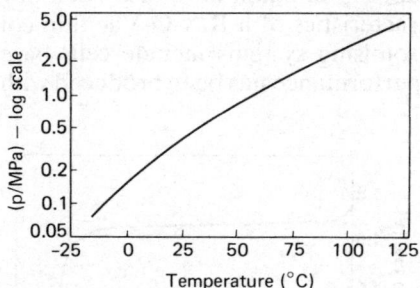

5.20 Internal pressure of lithium–sulphur dioxide cells as a function of temperature. (By permission of Elsevier Sequoia S.A.)

internal pressure decreases as the cell is discharged. However the pressure increases to 3 MPa (30 atm) at 100°C as shown in Fig. 5.20, so that the engineering problems associated with designing and fabricating a safe container and seals are severe. The cells have a spiral wound construction as illustrated schematically in Fig. 5.21. The lithium foil strip is separated from

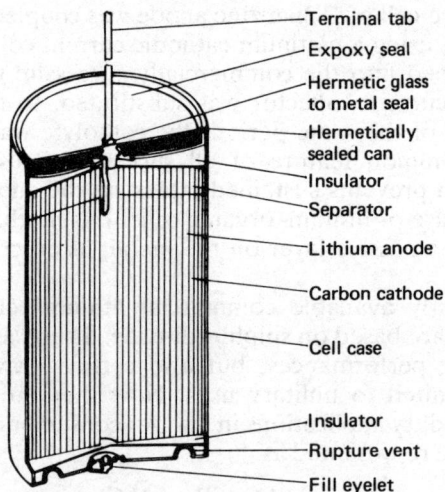

5.21 Lithium–sulphur dioxide spiral cell. (By permission of Elsevier Sequoia S.A.)

the cathodic current collector, a carbon black/Teflon mix, by a thin microporous polypropylene separator. A nickel plated steel case encloses the spiral and acts as the negative terminal. The positive connector is attached to the cathode via a central tantalum post enclosed in a glass-to-metal seal. The can is hermetically sealed and the liquid phase is injected through a filling eyelet or 'fillpoint' which is then welded shut. A safety vent is located at the bottom of the cell to prevent rupture of the outer case if the internal pressure were to rise above the safety limit.

Lithium–sulphur dioxide cells are characterised by energy densities of up to 330 Wh kg^{-1} (525 Wh dm^{-3}) which are four times as high as the best zinc and magnesium cell values. Because of the low internal resistance of the spiral wound cells, the discharge voltage of 2.8–2.9 V is hardly affected by the rate of discharge. In Fig. 5.22 typical discharge curves of a D-size cell (internal resistance 0.8 Ω) under various loads are shown. It can be seen that the discharge profiles are flat, with negligible voltage delay. Even at temperatures as low as −40°C, the voltage delay is minimal. The cell can operate at temperatures down to −54°C, where no conventional system is active, and up to +70°C. The shelf life of the cells is very long: it is claimed that less than 10% of the capacity is lost on storage over a five year period at room temperature.

At present, lithium–sulphur dioxide cells are manufactured in cylindrical form with capacities ranging from 0.45 to 30 Ah. They are classified as standard or high rate systems. The former are constructed to operate over a wide scale of load and temperature conditions. The latter are designed to deliver a high energy output at low operating temperatures. Applications in the military area take advantage of the high power density and unrestrictive operating conditions, and include providing power for portable radio transceivers, night vision equipment, sonobuoys, missiles and 'artillery delivered' devices*. Emergency systems, alarms, aircraft emergency locators, etc., are typical examples of civilian uses.

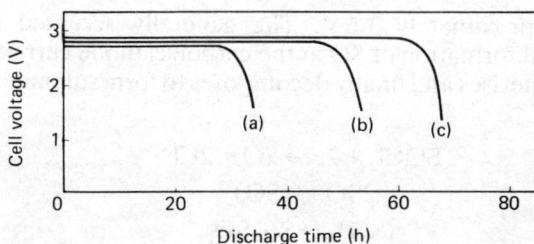

5.22 Discharge curves for D-size lithium–sulphur dioxide cells at ambient temperature: (a) 270 mA; (b) 180 mA; (c) 140 mA. (By permission of Elsevier Sequoia S.A.)

---

* Tactical nuclear weapons are now produced as 20 cm diameter shells.

Lithium–sulphur dioxide cells are characterised by high energy density, high power density, good voltage regulation, exceptional low temperature performance and superior shelf-life. The real problem is that if the cells are abused they may explode or vent a highly toxic gas. Exposure of the cell to high temperatures, either external or internally generated by prolonged short circuit conditions, may raise the internal pressure to dangerous levels. Further, after storage at 60°C for extended periods, some corrosion of the glass-to-metal seal has been observed. As usage changes from predominantly military to civilian applications, even more stringent safety standards become essential. In particular, improvements in seal and vent technology must be made. A separate problem which must be solved before the use of this system becomes widespread, concerns the safe disposal of spent cells.

(c) Liquid cathodes

A number of inorganic molecules such as thionyl chloride ($SOCl_2$), sulphuryl choride (SO_2Cl_2) and phosphoryl chloride ($POCl_3$) have been found capable of acting both as solvent and as cathodic reagent in lithium cells. Such materials are liquid over a wide temperature range, and can dissolve considerable quantities of salts such as $LiAlCl_4$ to give conductive solutions, as noted above in Table 5.4. Moreover, they show excellent kinetic stability towards lithium, due to the formation of passivating layers. Liquid cathode cells have many similarities to the lithium–sulphur dioxide system discussed above, except that no organic component is present. From Table 5.11, which lists the physical properties of the three materials, it is seen that thionyl chloride has the widest liquid range: commercial interest has been focussed mainly on this system.

The Li–$SOCl_2$ cell may be represented as

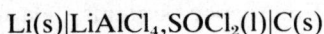

$$Li(s)|LiAlCl_4,SOCl_2(l)|C(s)$$

with the main cell reaction

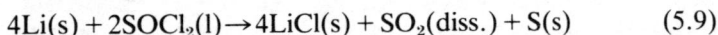

$$4Li(s) + 2SOCl_2(l) \rightarrow 4LiCl(s) + SO_2(diss.) + S(s) \qquad (5.9)$$

The OCV is approximately 3.6 V. The generally accepted mechanism involves the initial formation of SO at the carbon cathode current collector. This molecule dimerises and finally decomposes to form sulphur and sulphur dioxide:

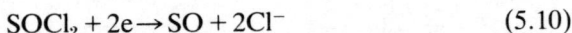

$$SOCl_2 + 2e \rightarrow SO + 2Cl^- \qquad (5.10)$$

$$2SO \rightarrow (SO)_2 \qquad (5.11)$$

$$(SO)_2 \rightarrow S + SO_2 \qquad (5.12)$$

The cell reaction thus leads to the formation of relatively insoluble lithium chloride at the cathode which under some circumstances may block the porous carbon current collector, leading to an increased internal resistance. Another product is SO_2, which may lead to an increasing internal pressure as the discharge continues. However in practice most of the SO_2 remains dissolved and the pressure build up is small. Further electrochemical reduction of SO_2 to form lithium dithionite may also occur. A number of

Table 5.11

Properties of liquid cathode materials

Name	Formula	Relative Molec. mass	m.p. /°C	b.p. /°C	Relative Permittivity	Viscosity /Nm⁻²s × 1000	Density /g cm⁻³
Phosphoryl Chloride	$POCl_3$	153.33	2	105	13.3 (22°C)	0.921 (25°C)	1.67 (25°C)
Sulphuryl Chloride	SO_2Cl_2	134.97	− 54.1	69.1	10 (22°C)	0.674 (25°C)	1.67 (25°C)
Thionyl Chloride	$SOCl_2$	118.97	−105	78.8	9.25 (20°C)	0.603 (25°C)	1.65 (25°C)

other possible reactions have been suggested, and as utilisation of $SOCl_2$ is much higher at low current drains, it would appear that the overall reaction is a function of the discharge rate.

Practical cells are mainly of the cylindrical or button type. The former may be designed with slightly different internal structures for high energy or high power applications, as shown schematically in Fig. 5.23. The high energy cell is a bobbin-type unit, while the high power cell has a spiral construction giving a large electrode area. The cathode current collector is again a porous carbon black/Teflon mix. Since $SOCl_2$ is a very aggressive chemical, there are special problems of material compatibility in cell fabrication. The cell case is constructed of nickel or nickel plated stainless steel and must be hermetically sealed to a helium leakage rate of $10^{-9} cm^3 s^{-1}$ or better.

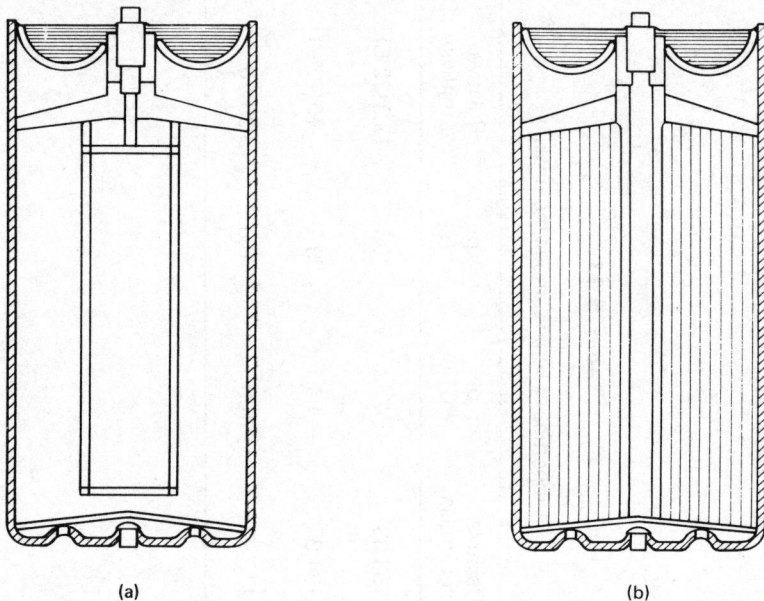

(a)   (b)

5.23 Schematic structure of D-size lithium–thionyl chloride cells: (a) high energy type; (b) high power type.

Normal plastics and rubbers are unstable in thionyl chloride and only fully fluorinated polyolefins such as Teflon can be used as insulators. Early manufacturing problems concerned with attack on glass-to-metal seals have been solved. Much larger cells with capacities up to 10,000 Ah and 40 A discharge capabilities have recently been tested, and a 1.2 MWh 30 V battery has been developed for military use.

At room temperature, discharge curves are exceptionally flat (Fig. 5.24), even at rates of up to 3 A for D-sized cells. As noted above, $SOCl_2$ utilisation is much higher at low current drains. Low rate cells manufactured by Mallory with a spiral configuration (4.45 cm $\times$ 25–38 cm electrodes) produced practical energy densities of 661 Wh kg^{-1} (1240 Wh dm^{-3}) at

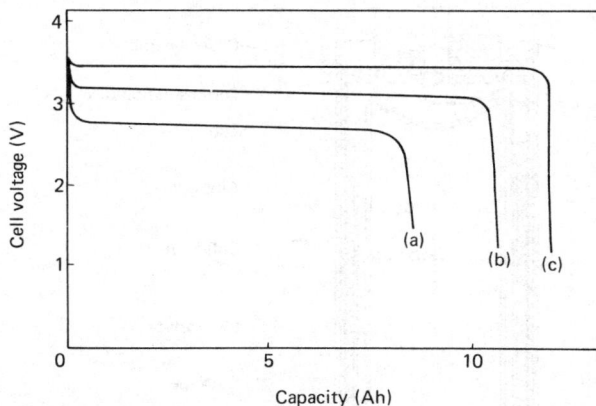

5.24 Discharge curves of D-size lithium–thionyl chloride cells at ambient temperature: (a) 3.0 A; (b) 1.0 A; (c) 0.1 A.

0.01 A. At temperatures of $-30°C$ and below (the working range for such cells is generally quoted as $-40°C$ to $+75°C$), some voltage delay is evident at higher current densities. Reasonably flat discharge curves are still developed for currents of 0.1 A and below (for D-sized cells), while 3 A can still be drawn without undue polarisation, although the cell capacity is much reduced. Shelf-life is excellent, with an estimated capacity loss of less than 0.5% per annum over a three year period.

Low rate cells are considered to be as safe as conventional batteries and have been used in cardiac pacemakers since 1974. The safety of high rate Li–$SOCl_2$ cells on the other hand still presents serious problems as there may be the danger of explosion for cells which have been short-circuited or reversed by forced discharge. Some manufacturers incorporate pressure vents in the cell casing. However according to one manufacturer, Tadiran, bobbin-type lithium-limited cells can be designed to withstand the most severe discharge/mechanical/heating tests. This is achieved (i) by increasing the heat dissipation of the system by swaging the lithium foil to the inside wall of the casing, and (ii) by limiting the reactive electrode area.

Many of the practical applications of lithium–thionyl chloride cells take advantage of their high energy density and favourable low temperature characteristics. Thus they are used in balloon and rocket borne meteorological radiosondes, emergency locating transmitters, underwater instrumentation, etc. Cells are also manufactured with special dimensions and terminals for direct mounting on printed circuit boards. Miniature (button) cells based on the Li–$SOCl_2$ system are used in a variety of implanted biotelemetry packages. The very large 40 A cells are used exclusively for military purposes – e.g. for driving torpedo motors.

Lithium–thionyl chloride cells are also constructed as reserve batteries for military use. A schematic diagram of the Tadiran system is shown in Fig. 5.25. The $LiAlCl_4$/$SOCl_2$ solution is contained in a sealed ampoule and the cell is activated by breaking this, either manually or by a spring loaded or pyrotechnic device.

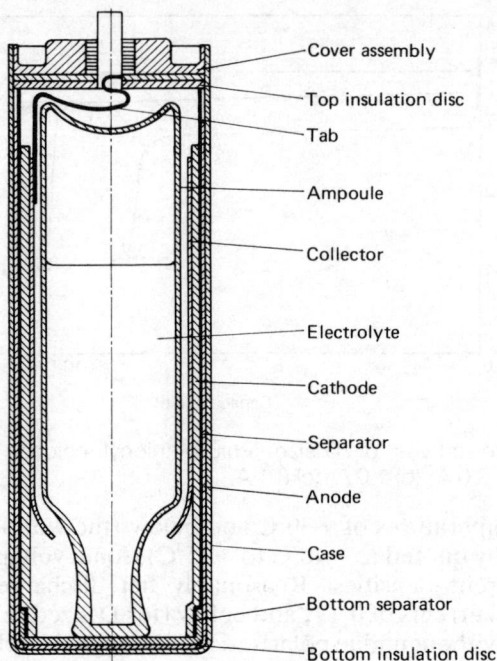

5.25   Cross-section of a lithium–thionyl chloride reserve cell. (By courtesy of Tadiran)

Calcium is being studied as a possible replacement for lithium in certain $SOCl_2$ cells.

Sulphuryl choride, when coupled with lithium, has an even higher density than thionyl chloride. The OCV of the cell

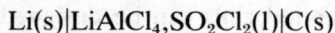

$$Li(s)|LiAlCl_4, SO_2Cl_2(l)|C(s)$$

is 3.91 V at 30°C. Manufacturing problems with such a cell are very similar to those of the Li–$SOCl_2$ system, so that it is expected that commercial developments of this cell will follow quite rapidly. Phosphoryl chloride suffers from a relatively high melting point (2°C) and a consequently restricted liquid range.

5.5 Secondary systems

Ambient temperature lithium cells are able to operate with the same energy and power characteristics as the successful high temperature sodium–sulphur and lithium–iron sulphide systems. If efficient rechargeable lithium–organic cells capable of a long cycle life could be developed, they would provide a promising alternative for traction and other applications where energy density is an important parameter. As discussed in Chapter 4, the general problem of devising a practical secondary cell is that of finding two compatible electrode systems, both of which show good reversibility. Considerable progress has been made in this search during the last four or five years, but there is still no commercial room temperature lithium cell capable of sustaining multicycle operation at even a medium charge/ discharge rate.

The lithium electrode

The existence of a passivating film on lithium surfaces in contact with an organic solvent-based electrolyte was discussed earlier (Section 5.3). Film formation ensures long shelf-life for lithium cells, but causes severe problems for good lithium electrode cyclability.

It is found that lithium can be plated with virtually 100% efficiency in a range of organic systems; however the plated lithium cannot be stripped quantitatively, especially if the cell has been allowed to stand for a period between plating and stripping. An explanation for this behaviour, advanced by Brummer of EIC Corporation, is illustrated in Fig. 5.26. The lithium is considered to be electrodeposited in granular form and the newly created surfaces react rapidly with components of the electrolyte, and this continues once the charging (plating) current has been switched off. Some lithium grains become partially undercut and others are completely isolated from the underlying lithium metal by an insulating film. Discharge (stripping) efficiency is therefore less than 100% and the residual isolated lithium grains affect the morphology of any subsequent replating. After a few cycles, the capacity or Ah-efficiency of the cycle falls to almost zero.

It was found that by altering the constituents of the electrolyte the nature of the passivating film could be modified to such an extent that the cycling behaviour was greatly improved. Attempts were therefore made to find an

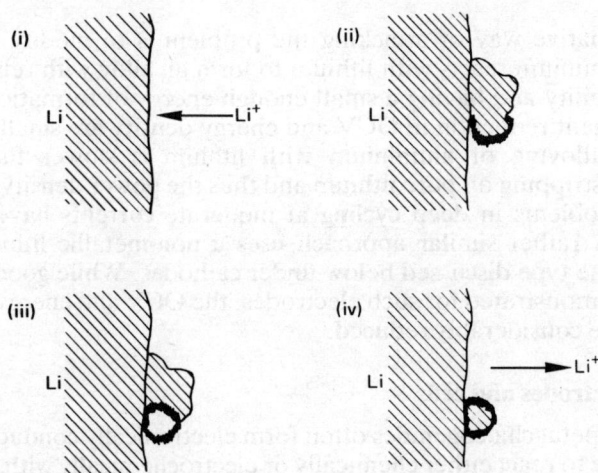

5.26 Mechanism for the isolation of plated lithium by an insulating film, as suggested by S.B. Brummer of EIC Corporation.

optimised electrolyte system which would result in the formation of 'desirable' films: i.e. films which would be impermeable to solvent and stabilise the metal, but which would remain conductive to lithium ions. One of the best electrolytes to date contains $LiAsF_6$ dissolved in carefully purified 2–methyltetrahydrofuran (2–Me–THF). This solvent was selected (a) because of the relatively low polarity of the C–O ether bond, and (b) because of the predicted effect of the methyl group on slowing the formation of ring-opened products by the lithium. The effect of changing the electrolyte constituents is shown in Fig. 5.27. However the exact reasons for the excellent behaviour of the $LiAsF_6$/2–Me–THF system are not fully understood. An experimental cell based on this electrolyte has sustained over one hundred deep cycles, but some loss in stripping efficiency is still evident when cells are allowed to stand for long periods at open circuit.

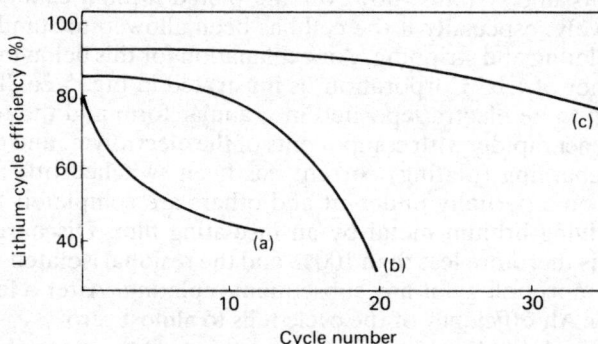

5.27 Effect of electrolyte composition on lithium cyclability, after results by V.R. Koch et al. of EIC Corporation: (a) $LiClO_4$, PC-purified electrolyte; (b) $LiAsF_6$, PC; (c) $LiAsF_6$, 2 Me–THF.

An alternative way of attacking the problem is to modify the lithium phase. Aluminium reacts with lithium to form an alloy with relatively high lithium mobility and having a small enough energy of formation such that the consequent reduction in OCV and energy density are small. However, alloying/dealloying of aluminium with lithium is slower than electro-deposition/stripping on pure lithium and thus the power density is reduced. Further, problems in deep cycling at moderate currents have still to be resolved. A rather similar approach uses a non-metallic lithium host as anode, of the type discussed below under cathodes. While good cyclability has been demonstrated for such electrodes, the OCV and energy and power densities are considerably reduced.

Positive electrodes and cells

Transition metal chalcogenides often form electronically conducting phases that are able to react either chemically or electrochemically with lithium in a reversible manner to form 'insertion' or 'intercalation' compounds: e.g.

$$xLi + TiS_2 \rightleftharpoons Li_xTiS_2 \tag{5.13}$$

or
$$xLi^+ + TiS_2 + xe \rightleftharpoons Li_xTiS_2 \qquad (5.14)$$

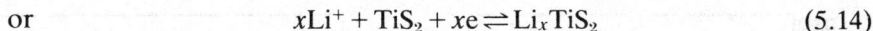

This behaviour is related to the structure of the 'host' material: in many cases the structure has layers or channels which can incorporate guest atoms so that a ternary phase can be formed with a minimum of structural perturbation. Thus an almost limitless number of charge/discharge cycles are possible without significant degradation of the structural or electrical properties of the host lattice. In many cases the mobility of lithium in the host lattice is reasonably good (e.g. the diffusion coefficient of lithium in TiS_2 is $2 \times 10^8\,cm^2\,s^{-1}$ at room temperature), so that concentration polarisation is not unacceptably high.

Layered structure dichalcogenides have been studied extensively. In Fig. 5.28 a schematic diagram of TiS_2 is shown. The titanium atoms are held between hexagonally close-packed sulphur atoms to form layers which are held together by weak van der Waals forces. Incorporation of lithium atoms between the layers produces a continuous and reversible variation of the

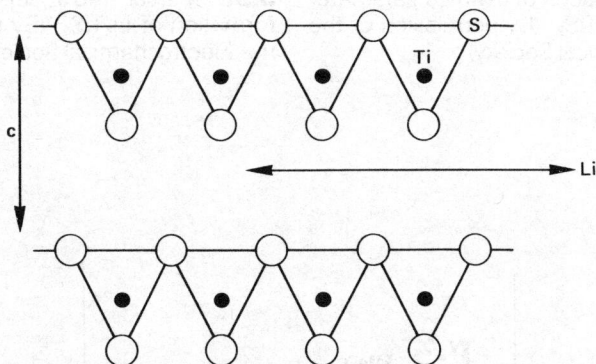

5.28 Schematic structure of TiS₂ layers.

c–lattice parameter as shown in Fig. 5.29, but no phase change. The electrode potential of Li_xTiS_2 with respect to lithium falls from about 2.4 V to 1.9 V as x rises from zero to one (Fig. 5.30). A wide range of possible reversible cathode materials has been studied. In Fig. 5.31 the approximate voltage ranges for reversible operation of a number of lithium/transition metal chalcogenides are given, and in Table 5.12 properties of some interesting systems are reported. However it should be noted that the lithium (anode) cyclability problem has limited the commercial development of such cells to the single example of the LiAl–TiS₂ cell.

This miniature cell was manufactured by Exxon in two sizes with capacities of 25 mAh and 90 mAh. It had a nominal voltage range of 1.5–2.1 V and was rated for a minimum of five cycles at 1 mA to 80% of nominal capacity. Due to limited commercial success, production was discontinued.

The original intention was to develop EV and load levelling batteries based on this system. However after considerable research it was found that the technical problems were too severe to proceed further at this time.

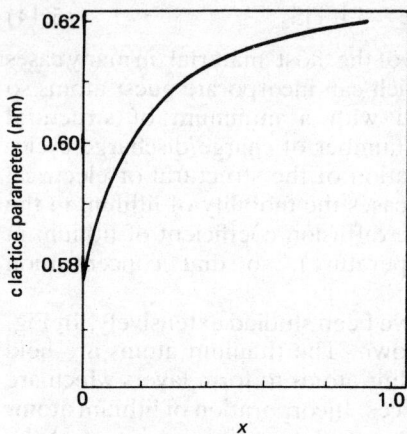

5.29 Variations of c-lattice parameter with x in Li_xTiS_2. (By permission of the Electrochemical Society)

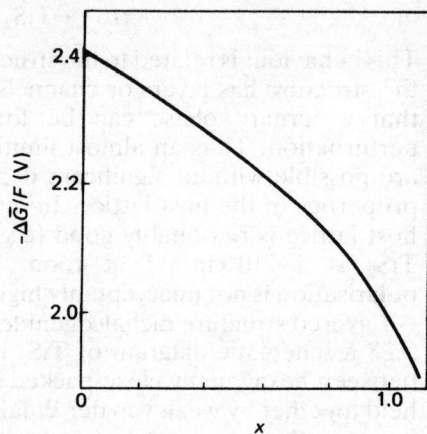

5.30 Partial molar free energy of formation of Li_xTiS_2. (By permission of the Electrochemical Society)

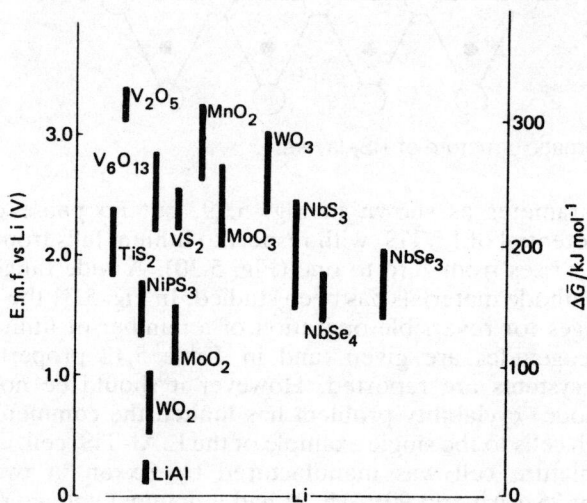

5.31 Voltage ranges for reversible operation of materials which may be coupled with lithium, after F.A. Trumbore. (By permission of Pure and Applied Chemistry).

Table 5.12

Selected reversible positive electrode materials in organic eletrolyte lithium cells

Compound	Structure	Reaction*	Potential Range vs. Li (V)	Theoretical Energy Density (Wh/kg)	Cyclability
TiS_2	layered, 2-D	$xLi + TiS_2 = Li_xTiS_2$ $0<x<1$	2.6–1.9	480	prolonged cycling at 80% depth
V_6O_{13}	rutile, 3-D	$xLi + V_6O_{13} = Li_xV_6O_{13}$ $0<x<7.9$	2.8–2.2	800	prolonged cycling in the $0<x<0.7$ composition range
$NbSe_3$	fibrous, 2-D	$xLi + NbSe_3 = Li_xNbSe_3$ $0<x<3$	2.2–1.2	440	over 200 deep cycles
Mo_8O_{23}	framework, 3-D	$xLi + Mo_8O_{23} = Li_xMo_8O_{23}$ $0<x<0.8$	2.8–2.0	1000	prolonged cycling at 40% depth
TiO_2	anatase, 3-D	$xLi + TiO_2 = Li_xTiO_2$ $0.2<x<0.6$	2.0–1.0	950	prolonged cycling in the $0.10<x<0.30$ composition

* The reactions are of the topochemical type, involving insertion in a host structure.

6 High temperature cells

6.1 Introduction

The primary requirements for large scale energy storage batteries, designed for applications such as vehicle traction, load levelling, etc., are high power and energy densities, coupled with low cost. For optimum performance and commercial success, electrode materials must therefore be selected which are very reactive and also abundant in nature. This in turn leads to the conclusion that lithium and sodium, being among the lightest, most electropositive and abundant of metals, should be ideal anode materials for such advanced battery systems. Because of their reactivity, these alkali metals cannot be used in conjunction with aqueous electrolytes. Nor can the non-aqueous solvent systems discussed in Chapter 5 be used: these have limited values of conductance so that the power levels of batteries based on them are severely restricted. Two further classes of electrolyte have therefore been exploited, namely fused salts and solid ionic conductors. Systems using fused salt electrolytes must be maintained at temperatures high enough to keep the electrolyte in a molten condition; batteries with solid electrolytes must be operated at temperatures which assure (a) a sufficiently high conductance of the electrolyte and (b) that in general the electrode phases are in a liquid condition to minimise electrode/electrolyte contact problems. One can therefore define a new group of power sources, which, requiring operating temperatures above ambient, are classified as **high temperature cells**. In Fig. 6.1 the range of specific power and energy densities offered by these batteries is compared with those of other advanced and traditional systems.

The energy content is seen to be sufficiently high to justify the present interest in this class of battery. However there are many problems concerning reliability, cycle life, safety and others connected with the advanced technology of these systems. These must first be solved before any prediction of their wide commercial acceptance can become proven. Up until now, production has been limited to laboratory models, commercial prototypes and in some cases, small scale pilot factory production. Nevertheless very considerable research and development programmes are being devoted to these batteries and it is considered likely that one or more systems will make a significant commercial impact in the near future.

The most notable of these systems, viz. the lithium–iron sulphide, and the sodium–sulphur batteries will be described in some detail in this Chapter. Some attention will also be given to interesting modifications of these basic

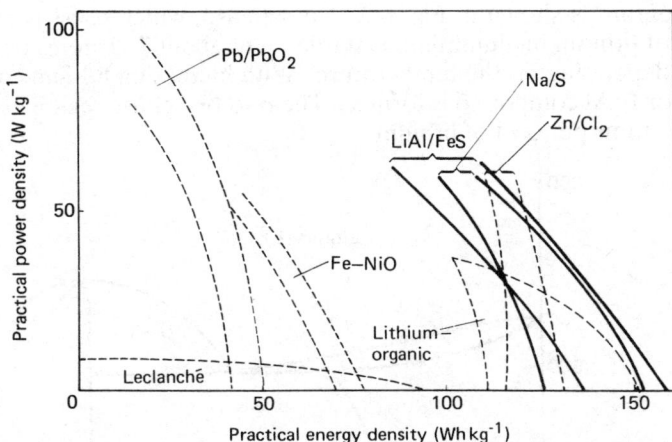

6.1 Power density-energy density curves for practical battery systems.

systems which, even if still in the early stages of development, appear promising in terms of performance or reliability. Related to the batteries surveyed above are the **thermal batteries** which are reserve primary systems activated, for example, by melting an appropriate salt mixture to form the electrolyte. A number of such systems, which are able to supply high currents for short periods, and have military, aerospace and emergency applications, will also be discussed. Before describing the construction and performance of individual systems, the basic properties of lithium and sodium electrodes and electrolytes commonly used in high temperature batteries will be briefly reviewed.

6.2 Negative electrodes

Lithium-based electrodes

The physical properties of lithium metal were given in Table 5.5. Despite its obvious attractions as an electrode material, there are severe practical problems associated with its use in liquid form at high temperatures. These are mainly related to the corrosion of supporting materials and containers, pressure build-up and the consequent safety implications. Such difficulties were experienced in the early development of lithium high temperature cells and led to the replacement of pure lithium by lithium alloys, which despite their lower thermodynamic potential, remained solid at the temperature of operation and were thus much easier to use. Recently, however, successful experiments have been carried out using low surface area, vertically pleated, stainless steel wicks in conjunction with molten lithium containing 1–2 atoms per cent of copper to promote wetting of the steel.

Of the various solid intermetallic lithium compounds which might be used in high temperature cells the Li–Al system has been most studied. The Li–Al phase diagram is shown in Fig. 6.2. An α-phase, which consists of a solid solution of lithium in aluminium is stable up to about 7–9 atoms per cent of lithium, depending on the temperature. With increasing lithium content, a β-phase or LiAl compound is formed. The α–β two phase region extends to about 46 atoms per cent of lithium.

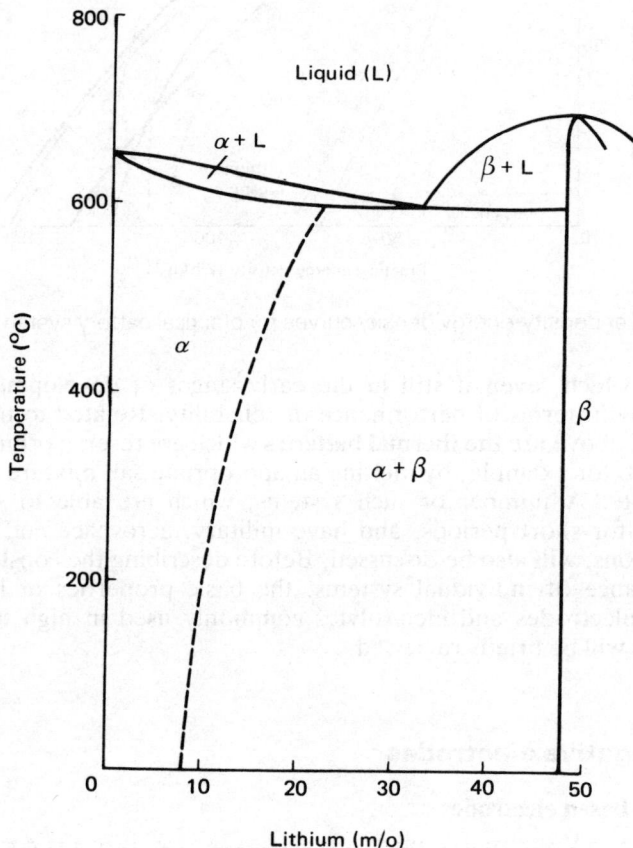

6.2   The lithium–aluminium phase diagram.

The e.m.f. of the lithium–aluminium system versus pure lithium in a LiI–KI–LiCl molten eutectic is shown in Fig. 6.3 as a function of temperature and composition. It can be seen that the e.m.f. remains constant (at about 300 mV more negative than pure lithium) in the range of stability of the β-phase ($\approx$7–47 atoms per cent of lithium), thus implying a constant lithium activity in the alloy surface. At concentrations greater than 47 atoms per cent, the lithium activity becomes strongly composition dependent.

Li–Al alloys may be prepared electrochemically – generally by coulometric deposition of lithium from a molten salt bath, or pyrometallurgically, by heating lithium and aluminium at a temperature above

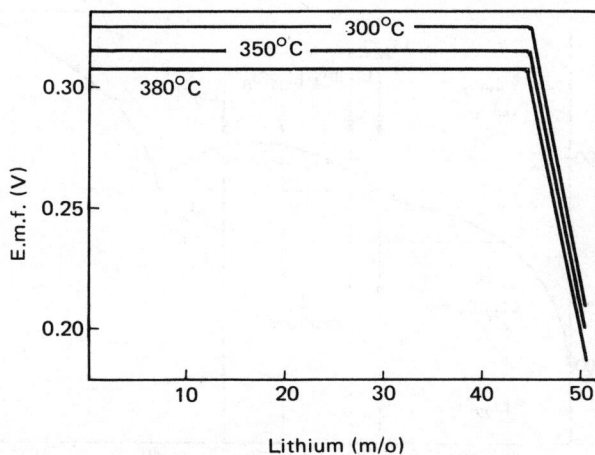

Figure axis labels: E.m.f. (V) on vertical axis (0.20, 0.25, 0.30); Lithium (m/o) on horizontal axis (10, 20, 30, 40, 50). Curves labelled 380°C, 350°C, 300°C.

6.3 E.m.f. of lithium–aluminium alloy as a function of temperature and composition. (By permission of the Electrochemical Society)

the alloy melting point of 720°C. Practical electrode configurations are constructed by

(i)   electrochemical deposition on a substrate of compressed aluminium fibres,
(ii) hot-pressing powdered mixtures of pyrometallurgical LiAl and electrolyte, or
(iii) loading pyrometallurgical LiAl powder into a porous nickel mass using vibratory techniques.

All these operations, and more generally, the assembly of the complete cell, must be performed in a controlled, water free atmosphere.

Li–Al electrodes behave well in molten salt electrolytes such as the LiCl–KCl eutectic, showing low polarisation and good reversibility, with flat charge/discharge characteristics, even under high current densities. Cycling behaviour may however be affected by progressive capacity losses. This problem may be partially solved by further modification of the alloy by the addition of a ternary component. Inclusion of approximately 4 per cent by weight of indium significantly increases capacity retention on cycling.

The Li–Si system has also been investigated as a possible high temperature solid lithium electrode. The lithium–silicon phase diagram, illustrated in Fig. 6.4, reveals five different compounds in the lithium-rich region. The existence of these compounds may be demonstrated electrochemically by the coulometric titration of a silicon electrode in a LiCl–KCl eutectic molten electrolyte, as shown in Fig. 6.5. Typical charge/discharge cycles of LiSi in the same electrolyte at 407°C (Fig. 6.6) show reversibility and good current efficiency, but also high polarisation during charge. This latter effect, possibly due to restricted lithium permeation within the short composition ranges of some of the phases, has so far restricted the use of LiSi alloys in practical high temperature rechargeable batteries.

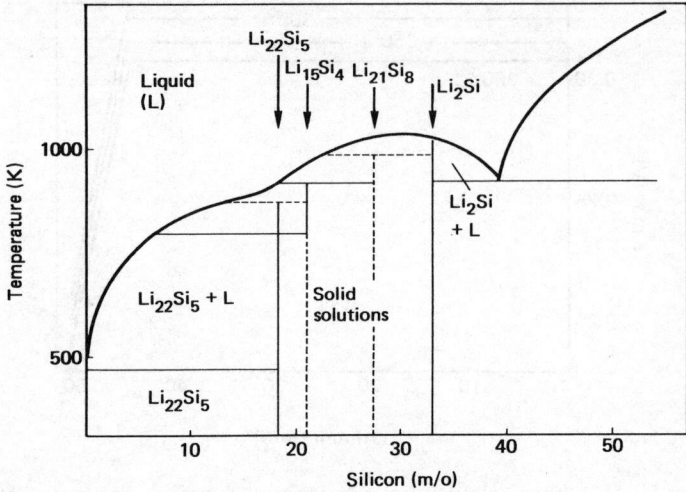

6.4 The lithium–silicon phase diagram. (By permission of the Electrochemical Society)

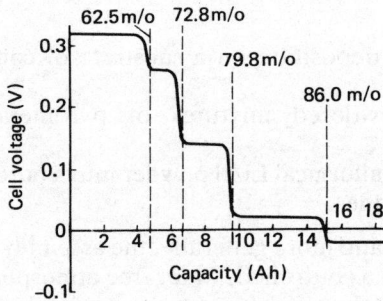

6.5 Coulometric titration of a silicon electrode with lithium using a lithium counter electrode and a LiCl–KCl eutectic at 680 K. Current density: 1 mA cm^{-2}. (By permission of the Electrochemical Society)

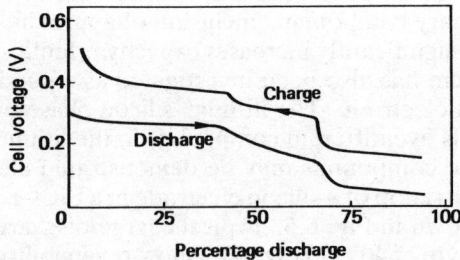

6.6 Typical charge/discharge characteristics of a lithium–silicon electrode in a LiCl–KCl eutectic at 680 K. Current density: 40 mA cm^{-2}. (By permission of the Electrochemical Society)

Sodium electrodes

Sodium is also a very reactive metal, and with a melting point even lower than that of lithium, presents in principle problems similar to those of lithium. However the fortunate discovery of ceramic and vitreous materials which show high stability to molten sodium together with good sodium ionic conductivity at high temperature has permitted the reliable fabrication of sodium-based cells. In the majority of prototype sodium high temperature cells, the liquid metal is housed in closed, shaped ceramic containers or within hollow glass fibres. In the others, the sodium is held in stainless steel cases or chambers. Special assemblies are constructed for vacuum-filling cells with molten sodium to prevent contact with oxygen or moisture.

6.3 Electrolytes

Lithium molten electrolytes

Molten single salts or molten eutectic mixtures are generally characterised by very high ionic conductivities ($> 100\,S\,m^{-1}$). This property makes them desirable media as battery electrolytes since it permits operation at high current densities with low iR drop, and hence without undue heat dissipation problems. On the other hand, molten alkali electrolytes require the maintenance of high temperatures. In addition to the necessity of energy input to heat the cell (at least initially) this requirement leads to a series of severe technical problems concerned with corrosion of cell housings, separators, ceramics for feedthrough insulators, current collectors and seals. The most common electrolyte used in lithium high temperature batteries is the LiCl–KCl eutectic which has a lower melting point than single lithium halides, but has the requisite high ionic conductance. However pure lithium dissolves in this electrolyte to an extent of about 0.13 m/o at 400°C which may lead to some self-discharge of cells. In addition, metallic lithium reacts with KCl to form potassium vapour with an equilibrium partial pressure of 0.05 kPa at 425°C. The properties of some lithium molten electrolytes are given in Table 6.1.

Molten chloroaluminates

Chloroaluminates, which are $AlCl_3$–MCl mixtures (where M is an alkali metal) offer a number of advantages as molten electrolytes, such as low melting point, high conductance and excellent stability (when contamination with oxygen and moisture is avoided). This melt can be considered as an acid-base system with the relative acidity or basicity being related to the chloride ion concentration. This in turn is governed by various equilibria: the predominant process near to the 1:1 $AlCl_3$/MCl ratio is

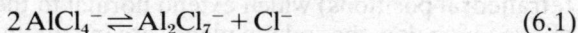

$$2\,AlCl_4^- \rightleftharpoons Al_2Cl_7^- + Cl^- \qquad (6.1)$$

$AlCl_3$–NaCl mixtures have been used in various high temperature batteries. The physical properties of the 1:1 system are summarised in Table 6.1.

Table 6.1

Physical properties of some lithium and sodium molten electrolytes*

Salt(s)	Composition	M.P. °C	Density g m^{-3}	Viscosity cP†	Conductance S m^{-1}
LiCl	—	610	1.490 (637°C)	1.377 (637°C)	585.4 (637°)
LiCl–KCl	58.5 m/o LiCl 41.5 m/o KCl	355	1.646 (447°C)	1.46 (617°C)	161.5 (457°C)
LiF–LiCl–LiBr	22 m/o LiF 21 m/o LiCl 47 m/o LiBr	445	2.190 (500°C)	—	—
NaCl–KCl	50 m/o NaCl 50 m/o KCl	685	1.571 (717°C)	1.58 (727°C)	239.6 (717°C)
NaCl–AlCl$_3$	50 m/o NaCl 50 m/o AlCl$_3$	155	1.691 (177°C)	2.645 (187°C)	46.2 (187°C)

* From G. Mamantov in 'Materials for Advanced Batteries', D.W. Murphy, J. Broadhead and B.C.H. Steele, Editors, Plenum Press, N.Y., London, 1980, p. 113.
† 1 cP = 0.001 kg m^{-1} s^{-1}

Sodium ß–aluminas

Sodium β–aluminas (often known as simply 'β–aluminas') are examples of **solid electrolytes**, i.e. compounds which permit fast ionic motion (here of sodium ions) within a solid lattice. While β–aluminas conduct reasonably well at room temperature (≈ 1 S m^{-1} for polycrystalline material) they are generally used at temperatures over 300°C where their conductance is greater than 10 S m^{-1}. 'Ambient' solid electrolytes and batteries based on these will be considered in the next chapter.

The most common phase is β–alumina itself, which has a formal composition of $Na_2O.11Al_2O_3$ but which in practice always contains an excess of Na_2O. Some of the physical properties of this material are given in Table 6.2. The high sodium ion mobility can be understood by examining the hexagonal layered β–alumina structure, shown schematically in Fig. 6.7. The structure is seen to contain planes having loose packed sodium and oxygen ions. Sodium planes are held 1.13 nm apart by four close-packed spinel blocks (oxygen layers with aluminium ions in octahedral and tetrahedral positions) which extend normal to the c-axis. The spinel blocks above and below the sodium planes are mirror images of each other and are separated by oxygen bridges of 0.48 nm. The sodium ions can move easily within the loosely packed planes, but not through the closely packed spinel

Table 6.2

Physical properties of β– and β''–aluminas

	β–alumina	β''–alumina
Melting point	2000°C	—
Density (at 25°C)	3.25 g cm^{-3}	—
a lattice constant	0.559 nm	0.559 nm
c lattice constant	2.253 nm	3.340 nm
Ionic conductance		
single crystal (at 25°C)	3.5 S m^{-1}	—
single crystal (at 300°C)	21.5 S m^{-1}	100 S m^{-1}
polycrystalline (at 25°C)	1.0 S m^{-1}	—
polycrystalline (at 300°C)	6.5 S m^{-1}	20 S m^{-1}
Electronic conductance (at 300°C)	~10^{-9} S m^{-1}	~10^{-9} S m^{-1}

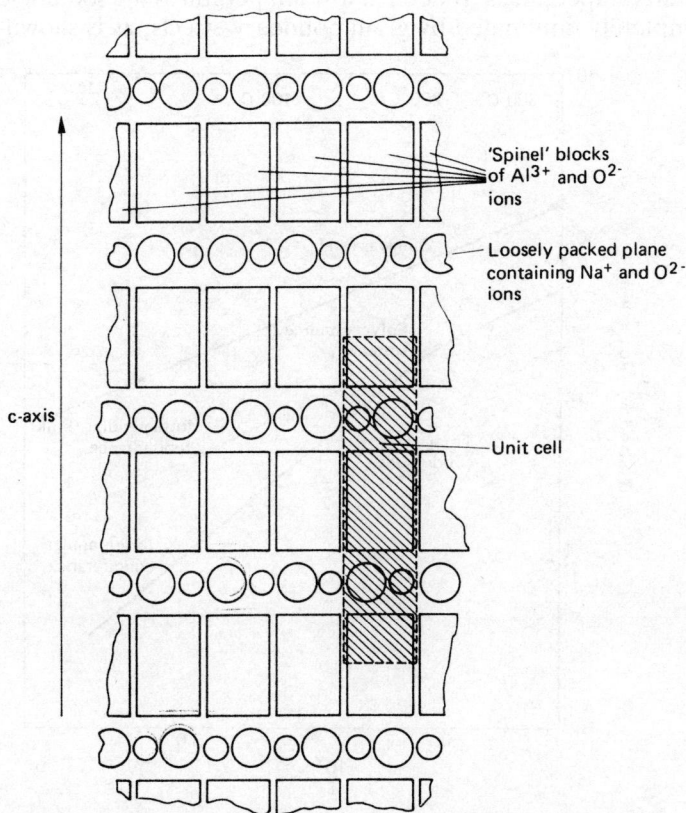

6.7   Schematic diagram of β–alumina showing loosely packed planes containing mobile sodium ions situated between spinel blocks of Al^{3+} and O^{2-} ions.

blocks. The structure therefore allows a two-dimensional diffusion of the sodium ions in directions perpendicular to the c-axis.

Another important phase in the Na_2O–Al_2O_3 system with a similar layered structure is the β'' form, of formal composition, $Na_2O(5.33Al_2O_3)$, whose basic properties are also summarised in Table 6.2. This phase is stabilised by small additions of MgO or Li_2O: the level of stabiliser addition affects the properties of the material, and the optimum composition in terms of conductance and mechanical strength appears to be around 0.8% Li_2O (with Na_2O at 9%). β''–alumina has a higher conductance than β–alumina and is therefore preferable as a battery electrolyte. However it is more sensitive to moisture and is thus more difficult to handle than β–alumina, so that a compromise mixture is generally used.

β–aluminas are almost always used as sintered polycrystalline masses because of the need to fabricate the electrolyte in particular shapes for battery application. The conductance of polycrystalline samples is lower than that of correctly oriented single crystals since the migration of sodium ions is affected by the tortuosity of the two-dimensional pathways and by grain boundary impedances. Indeed at low temperatures the sodium ion mobility is completely dominated by grain boundary effects, as is shown in Fig. 6.8

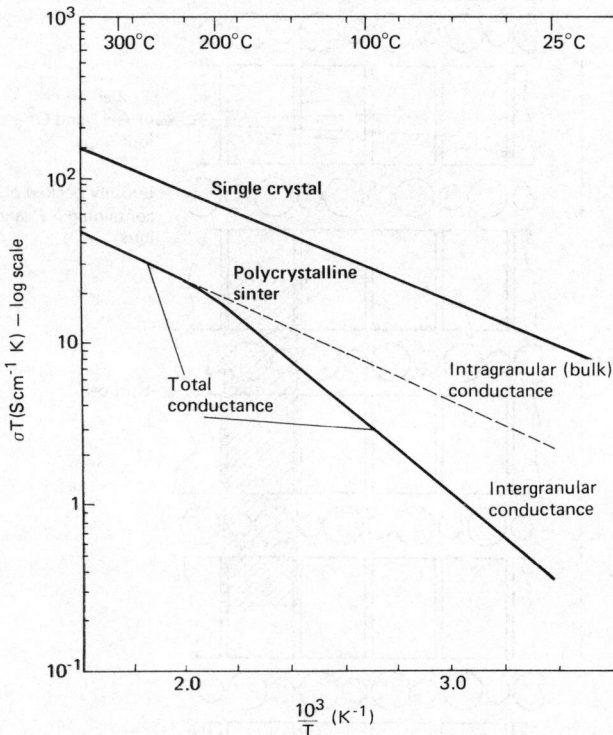

6.8 Arrhenius plots for single crystal and polycrystalline β–alumina. The overall conductance of the latter is dominated by the inter-granular impedance at low temperatures. (By permission of the Institute of Physics)

where the general trends in the conductivity of polycrystalline and single crystal β–aluminas as functions of temperature are compared. The presence of grain boundary impedances may give rise to intense localised electric fields within a polycrystalline sample during passage of current, and these fields may be responsible for progressive degradation of the electrolyte on prolonged cell operation.

Just as the electrical properties are clearly related to the phase composition and microstructure of the electrolyte, so are its key mechanical properties such as durability, resistance to thermal and mechanical shock, etc. Major research and development programmes are under way in France, Germany, Japan, the UK, and the USA to optimise electrical and mechanical properties and to fabricate the electrolyte in a suitable form. The most common configuration is of sintered, polycrystalline material in the form of a closed tube generally of diameter 2–3 cm, length 20–70 cm and wall thickness 1–2 mm. Some commercial battery electrolyte tubes are shown in Fig. 6.9.

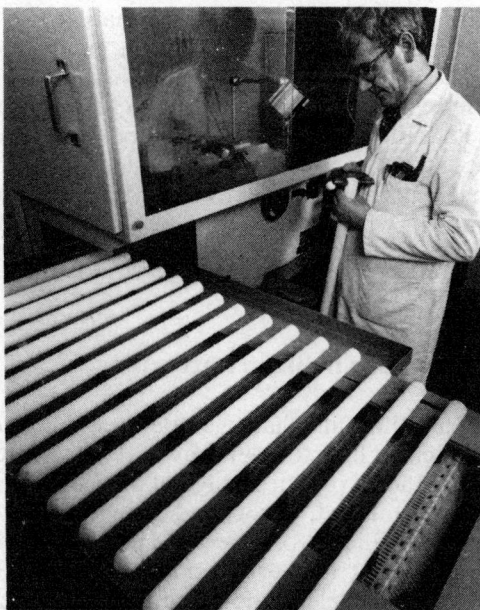

6.9 Checking tolerances of β–alumina electrolyte tubes in the quality control department of Chloride Silent Power Ltd.

Other high temperature solid electrolytes

Other crystalline sodium ion conductors, the so-called NASICON compounds of general formula $Na_{1+x}Si_xZr_2P_{3-x}O_{12}$ for $1.8 < x < 2.4$ have been investigated as possible alternatives to the β–aluminas. The advantage of this group of materials is their comparative case of preparation.

Lithium ions exhibit fast ion conduction in mixed Li–Na β–alumina, but this electrolyte decomposes in the presence of molten lithium. A considerable research effort is being mounted to develop a suitable lithium-compatible material. The most promising materials seem to be $Li_{1.8}N_{0.4}Cl_{0.6}$ and $Li_{3.6}P_{0.4}Si_{0.6}O_4$. The latter, in addition to being stable in the presence of lithium, is insensitive to a moist atmosphere.

Glass electrolytes

Sodium borate glasses which show some sodium ion conductivity ($\approx 4 \times 10^{-3} S\,m^{-1}$ at 300°C) have been known for many years. Recently a successful sodium glass electrolyte has been developed using bundles of thousands of hollow glass fibres. This configuration combines a very high superficial area with a wall thickness of some $10\,\mu m$, so that the total conductance of the electrolyte is high, despite the relatively high specific resistance.

From a technological point of view there are many advantages in developing highly conducting vitreous materials for battery electrolytes. First, the effect of grain boundaries can often be eliminated and, in addition, it is often easier and cheaper to fabricate glass into the form required for the final battery system. A wide range of vitreous lithium conductors has been studied, in particular lithium-rich phosphate, borate and aluminate systems. Some of these are stable in contact with lithium, but no commercial cell has yet been reported using such an electrolyte. A range of sodium conducting glasses is also known, with some Na_2S–SiS_2 glasses having as high an ionic conductance as the β–aluminas. Again use of these glasses in commercial sodium-based cells has yet to be reported.

6.4   High temperature lithium batteries

Lithium–halogen and lithium–chalcogen batteries have a certain historical importance in the field of high temperature batteries, since they were among the first systems to reach a practical stage of development. However severe problems were encountered, associated with the corrosion of the cell housing, seals and ancilliary components and thus with the stability and safety of the cells, so that further development of these systems has now practically ceased. Nevertheless, the results obtained drew attention to the energetic possibilities of lithium high temperature cells in general, and thus stimulated further research on alternative, stable lithium systems. It therefore seems useful to outline briefly the basic features of representative lithium–halogen and lithium–chalcogen batteries.

Lithium–chlorine batteries

The cell, shown schematically in Fig. 6.10, has the form

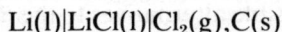

$$Li(l)|LiCl(l)|Cl_2(g),C(s)$$

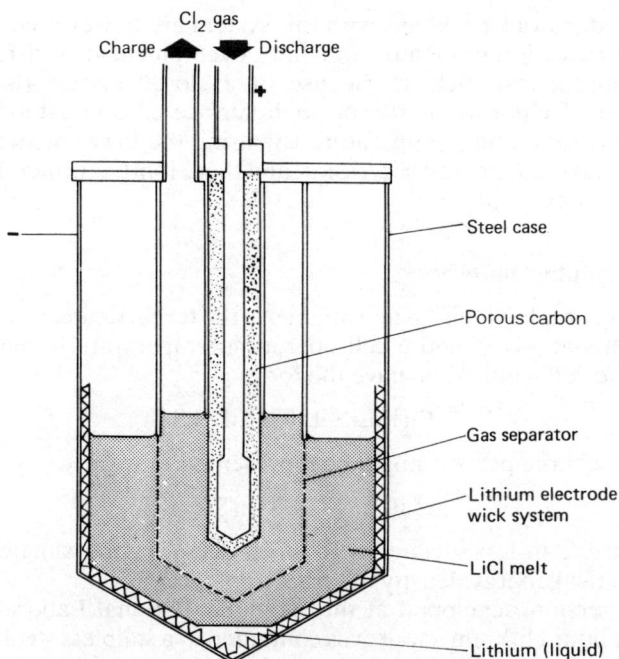

6.10 Schematic diagram of a vertical experimental lithium–chlorine cell.

The liquid lithium anode is held in a stainless steel case or nickel fibre pleated wick, and chlorine gas is fed under pressure from external storage to a porous carbon current collector. The two electrodes are separated by a molten lithium chloride electrolyte at 650°C. The products formed during recharge (i.e. liquid lithium and gaseous chlorine) must be liberated in such a way as to be easily separated and stored, in order to ensure proper cycling behaviour. This is achieved by using separator screens or special electrode structures which have porous, electronically insulating layers on the sides facing the electrolyte. When wetted by the electrolyte these layers form a seal which prevents or at least limits the escape of the products. The overall cell reaction is

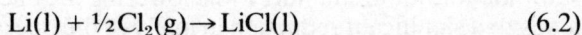

$$Li(l) + \tfrac{1}{2}Cl_2(g) \rightarrow LiCl(l) \tag{6.2}$$

with which is associated an e.m.f. of 3.46 V and a theoretical energy density of 2.18 kWh kg^{-1} at the temperature of operation. However the practical realisation of this very high energy is limited by a number of factors. Despite initial power densities of up to 40 W cm^{-3}, the available energy is reduced by the necessity of (i) maintaining the cell at 650°C and (ii) pumping the chlorine gas under pressure into the porous positive electrode. In addition, the 'valve' layers on the electrodes, described above, have a significant resistance which causes increased iR drop and thus depresses cell performance.

The most serious problems with this system are however concerned with corrosion of cell components and the development of satisfactory seals. Other approaches, such as the use of a closed system (based on the absorption of chlorine *in situ* on high surface carbon cathodes) and the reduction of operating temperature (by using the lower melting LiCl–KCl eutectic), have not proved successful, and there is little further development in progress at present.

Lithium–sulphur batteries

A typical example of this system might use the ternary eutectic LiF–LiCl–LiI which melts at 341°C and a cell operating temperature in the range 350–400°C. The cell would then have the form

$$Li(l)|LiF-LiCl-LiI(l)|S(l)$$

and the discharge process might be represented as

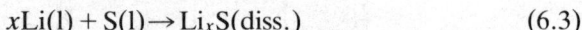

$$xLi(l) + S(l) \rightarrow Li_xS(diss.) \tag{6.3}$$

The average e.m.f. associated with this process is approximately 2.3 V and the theoretical energy density is $2.6\,kWh\,kg^{-1}$ at 350°C.

In the version developed at the Argonne National Laboratories in the USA, the liquid lithium anode was contained in a stainless steel cup and was separated from the liquid sulphur cathode by a sheet of zirconia cloth. However the all liquid system presented difficult practical problems, and a new type of cell using an immobilised electrolyte was proposed, where a more effective separation between the electrodes could be achieved. The electrolyte consisted of a rigid 'paste' of $LiAlO_2$ filler and LiF–LiCl–LiI eutectic in a 1:1 ratio. The first example of this type of cell had a tellurium cathode and was able to provide an e.m.f. of 1.7 V at 480°C and a practical energy density of $150\,Wh\,kg^{-1}$. However once again problems of corrosion and thermal stress in cell components reduced the cycle life to a level below that acceptable for a practical storage battery.

Lithium alloy–metal sulphide high temperature batteries

The substitution of a solid alloy for liquid lithium and a transition metal sulphide for liquid sulphur, while lowering the energy density of the system, leads to a significant reduction in many of the operating problems associated with the early lithium–chalcogen cells, and to easier fabrication of practical batteries. In particular, problems connected with the high vapour pressure of sulphur at the cell operating temperature (77 kPa at 425°C) and sulphur-displacement processes are eliminated, and the characteristic low polarisation and good reversibility of the lithium alloy electrodes can be exploited. The theoretical energy density is reduced from $2600\,Wh\,kg^{-1}$ to $400\,Wh\,kg^{-1}$, depending on the exact chemical composition of the cell. A wide variety of sulphides have been considered for the positive electrode, including those of Co, Cr, Cu, Mn, Ni, Ti and V, but FeS and FeS_2 have been selected for cell development on the basis of performance, cost and availability. Commercial batteries are being developed for both EV and load levelling applications mainly at the Argonne National Laboratory – in association with a group of industrial battery companies, Catalyst Research Corporation, Eagle-Picher and Gould in the USA, by Rockwell International, Sandia and General Motors, also in the USA, and by Varta in Germany. At the present time, full scale 40 kWh EV batteries are being tested and 5 MWh storage batteries are projected for load levelling tests in 1983.

The cell being developed for load levelling is denoted as:

$$LiAl(s)|LiCl-KCl(l)|FeS(s)$$

The cell reaction involves a two electron process and may be written formally as

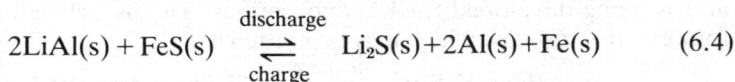

$$2LiAl(s) + FeS(s) \underset{charge}{\overset{discharge}{\rightleftharpoons}} Li_2S(s)+2Al(s)+Fe(s) \qquad (6.4)$$

The e.m.f. of this reaction is 1.33 V and the corresponding theoretical energy density is 460 Wh kg^{-1}. In reality the actual mechanism of the cell reaction is very complex and not yet fully understood: it may involve as many as six electrochemical steps and four chemical reactions, and has a different sequence depending on whether the surface (i.e. the area in contact with the electrolyte) or the interior of the FeS particles is involved. The discharge reactions are thought to occur in the following sequence:

Surface: $6LiAl + 26FeS + 6KCl \rightarrow LiK_6Fe_{24}S_{26}Cl + 2Fe + 5LiCl + 6Al$
$$(6.5)$$

Interior: $2LiAl + 2FeS \rightarrow Li_2FeS_2 + Fe + 2Al \qquad (6.6)$

$$2LiAl + Li_2FeS_2 \rightarrow 2Li_2S + Fe + 2Al \qquad (6.7)$$

Surface: $46LiAl+LiK_6Fe_{24}S_{26}Cl+5LiCl \rightarrow 26Li_2S+24Fe+6KCl+46Al$
$$(6.8)$$

The charging process is simpler and involves mainly the reactions

$$Fe + 2Al + 2Li_2S \rightarrow Li_2FeS_2 + 2LiAl \qquad (6.9)$$

and

$$Fe + 2Al + Li_2FeS_2 \rightarrow 2FeS + 2LiAl \qquad (6.10)$$

The formation of the compound $LiK_6Fe_{24}S_{26}Cl$ (which is also known as a mineral occurring in meteorites) is responsible for a characteristic swelling effect in the FeS cathode during discharge. This effect is less serious at high temperatures ($\approx 500°C$) and may be reduced by incorporating Cu_2S in the cathode mix, or by employing a potassium-free melt such as the LiF–LiCl–LiBr eutectic. Typical charge/discharge curves are shown for this system in Fig. 6.11.

The cell

$$LiAl(s)|LiCl-KCl(l)|FeS_2(s)$$

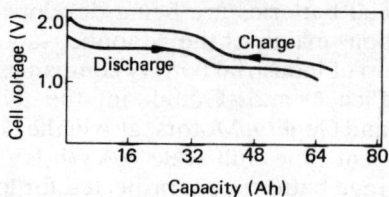

6.11   Charge/discharge characteristics for a Li/FeS cell at 450°C. Battery assembled in uncharged state. Current: 5 A. (By permission of Academic Press)

has a higher e.m.f. (1.76 V) and theoretical energy density (650 Wh kg^{-1}) and is being developed for EV applications. The overall cell reaction is a four electron process which may be written as:

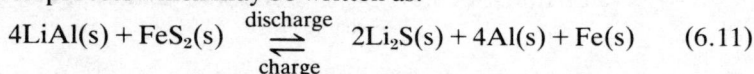

$$4LiAl(s) + FeS_2(s) \underset{charge}{\overset{discharge}{\rightleftharpoons}} 2Li_2S(s) + 4Al(s) + Fe(s) \qquad (6.11)$$

Again the actual cell reactions are very complex and the compositions of some of the ternary intermediates are only known approximately. The discharge mechanism is considered to proceed in three main steps:

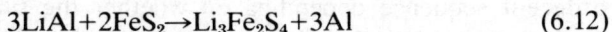

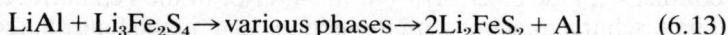

$$3LiAl + 2FeS_2 \rightarrow Li_3Fe_2S_4 + 3Al \qquad (6.12)$$

$$LiAl + Li_3Fe_2S_4 \rightarrow \text{various phases} \rightarrow 2Li_2FeS_2 + Al \qquad (6.13)$$

$$2LiAl + Li_2FeS_2 \rightarrow 2Li_2S + Fe + 2Al \qquad (6.14)$$

Traces of potassium compounds are also found due to reactions with the molten electrolyte. The sequence of phase changes during the charging process are even more complex. Typical charge/discharge curves are shown in Fig. 6.12. In practice, FeS$_2$-based cells tend not to be used in storage systems because of thermal decomposition of FeS$_2$. They are normally designed as high power primary devices and operate only on the upper voltage plateau, where polarisation is low.

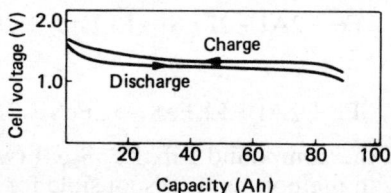

6.12   Charge/discharge characteristics for a Li/FeS$_2$ cell at 450 °C. Battery assembled in uncharged state. Current: 5 A. (By permission of Academic Press)

Cell materials

As with all high temperature batteries, the materials problem in the lithium alloy–metal sulphide system is one of trying to develop low cost components which are able to withstand the hostile environment of the cell. The choice of insulators and separators is greatly restricted by the high stability of Li_2O which excludes use of the common ceramics such as Al_2O_3 and SiO_2 for thermodynamic reasons. Other materials have been found unsuitable because of impurity content, formation of conductive surface layers, etc. The only effective stable fabrics so far reported have been made of BN (boron nitride) or Y_2O_3 (yttria). The former, although very expensive, is some six times cheaper than the latter. BN fabrics have been used as separators for most of the Argonne National Laboratory development programme. Rockwell (Atomics International Division) have been investigating rigid porous separators of Si_3N_4 or AlN but have encountered considerable problems connected with iR losses due to tortuosity and the concentration of impurities at grain boundaries during the sintering process. They have had more success with an inexpensive separator consisting of BN spacers and a packed ceramic powder of AlN or Y_2O_3. Powder separators, such as MgO are also currently under test. Cell housings and current collectors are constructed of low-carbon steel where possible, but more expensive current collectors based on molybdenum alloys have sometimes proved necessary. The electrodes consist of a porous bed of the solid active material located in an expanded metal or 'honeycomb' current collecting core, and flooded with molten salt. They are surrounded by a fine porous particle retainer or screen which confines the active material within the electrode structure. Electrodes may be formed in either the charged or discharged state.

Cell design and performance

The earliest prototype cells were cylindrical in form. A central metal sulphide electrode was imbedded in a molybdenum mesh current collector and this electrode was surrounded first by a layer of ZrO_2 felt, acting as particle retainer and then by a BN fabric separator. Upper and lower LiAl electrodes used the cell housing as current collector. Early cells of this design produced by the Argonne National Laboratory, called 'bicells' because of the double negative electrode, had typical practical energy densities in the range 100–140 Wh kg^{-1}, but poor cycle life and capacity retention. Better performance has been reported more recently for cylindrical cells developed by General Motors Research Laboratories, which are capable of more than 650 deep cycles. However the main thrust of lithium–iron sulphide cell design is now directed towards the production of vertically oriented cells with rectangular cross-section (Fig. 6.13). Again the central positive electrode is sandwiched between two negative electrodes which are in contact with the walls of the cell casing. The positive lead passes through an insulating compression seal in the top of the cell. Fine metal screens and/or zirconia or yttria felt are used as particle retainers and the positive mixture is wrapped in boron nitride fabric, which acts as separator. Typical Argonne National Laboratory cells are 13 cm × 13 cm or 13 cm × 18 cm and 1–3 cm thick, with theoretical capacities of 120–150 Ah. FeS-based cells have been shown to operate successfully to over 100 deep cycles, but with poor specific power; FeS_2-based cells on the other hand show peak specific power of more

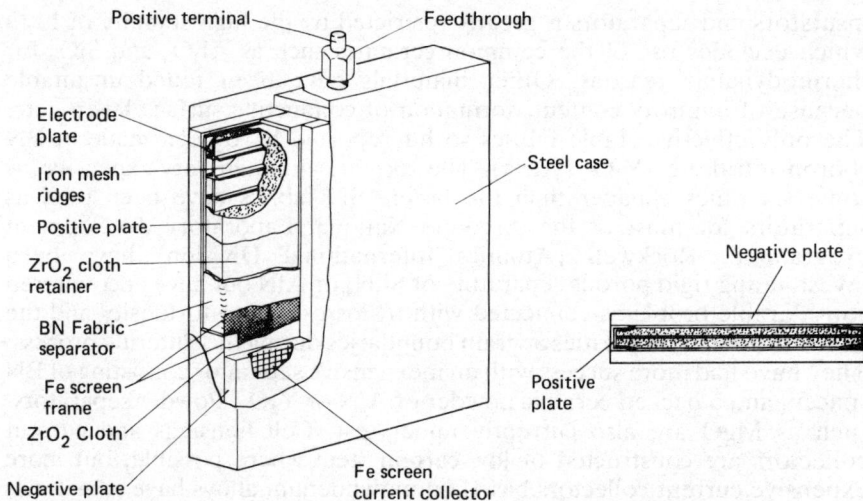

6.13 Rectangular (prismatic) lithium–iron sulphide battery. (By permission of Academic Press)

than $170\,W\,kg^{-1}$ and practical specific energies of over $100\,Wh\,kg^{-1}$, but poor cycle life. The Eagle-Picher FeS cell is 4 cm thick, but has a multiplate design with three positive and four negative electrodes (in parallel) in the one housing. This unit is able to develop a specific power of $95\,W\,kg^{-1}$. A 40 kWh prototype EV battery has been constructed, consisting of two 20 kWh modules (Fig. 6.14). Each module contains sixty multiplate cells housed in a thermally insulated casing equipped with a vacuum insulation annulus to minimise heat losses. Other development models based on this type of cell are also planned. All of the above cells have used the expensive BN fabric separators. As discussed above, Rockwell have investigated rigid and powdered ceramic separators. They have reported success with the latter in 2.5 kWh cells operating with the $Li_{3.5}Si$–FeS system, using AlN powder. Gould and Eagle-Picher are considering replacing BN fabric with felt, which is considerably cheaper.

Despite the considerable advances in the lithium alloy–metal sulphide battery made in recent years, a number of difficulties remain to be resolved before a significant commercial impact can be expected. In particular a solution must be found to the problem of capacity losses on cycling. These reduce the typical cycle life of a LiAl/FeS cell to around 120 deep cycles. Studies are being made into various possible side reactions which may occur during charging such as sulphur production or the formation of species involving iron and chloride. The need to use costly materials for ancilliary cell components is a serious commercial handicap, and the search for cheaper alternatives must continue. (The BN fabric can contribute up to 75% of the cost of cell materials.) The FeS-based system was selected for load levelling applications primarily on cost grounds, since relatively inexpensive iron current collectors can be used. LiAl–FeS cells perform well

6.14   20 kWh lithium–aluminium/iron sulphide battery. (By courtesy of Argonne National Laboratory)

at current densities up to $0.1\,A\,cm^{-2}$ but show a decrease in capacity at higher rates. On the other hand, FeS_2-based cells are suitable for EV use because of their greater available energy and power. However they require the more corrosion resistant (and thus more expensive) molybdenum current collectors.

If an acceptable cycle life and reliability can be achieved at a reasonable cost, successful penetration of the EV market is to be expected. Safety tests have shown that sudden rupture of the cells in air did not lead to explosions or fire, and that there was little escape of molten electrolyte. Early applications are likely to be in fork-lift trucks, urban delivery vans, postal vans, etc. A proposal has also been made to use FeS_2-based batteries in submarines. Tests are proposed for the BEST facility in 1983 to examine the technical and commercial feasibility of the LiAl–FeS battery in the large scale stationary energy storage role.

A LiAl high temperature battery using LiCl–KCl eutectic and a carbon/ $TeCl_4$ positive is being developed in the USA by ESB Corporation for EV use.

A calcium analogue

At this point mention might be made of an analogous cell system which uses calcium in the place of lithium. The development of calcium–iron sulphide cells is now a practical proposition, following the discovery of suitable calcium alloys, such as Ca_2Si and the ternary Ca_xMg_2Si.

Prototype cells of the form

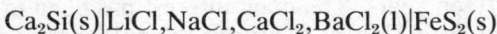

$$Ca_2Si(s)|LiCl,NaCl,CaCl_2,BaCl_2(l)|FeS_2(s)$$

have been produced with rectangular (prismatic) configurations, similar to those used for LiAl/FeS$_2$ cells. Again, BN fabric separators and molybdenum positive current collectors are used. The quaternary electrolyte has a lower conductivity than the LiCl–KCl eutectic but shows a lower solubility to the CaS reaction product. The Ca$_2$Si–FeS$_2$ cell operates at 450–500°C with an OCV of 2.0 V and a theoretical energy density of 790 Wh kg^{-1}. The discharge reaction

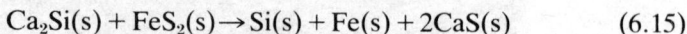

$$Ca_2Si(s) + FeS_2(s) \rightarrow Si(s) + Fe(s) + 2CaS(s) \qquad (6.15)$$

proceeds through a series of steps as shown in Fig. 6.15.

Practical energy densities of round 90 Wh kg^{-1} and a cycle life of over 100 cycles have been reported. Further studies are in progress to evaluate the utilisation of the Ca$_2$Si electrodes and the degradation of the BN separator.

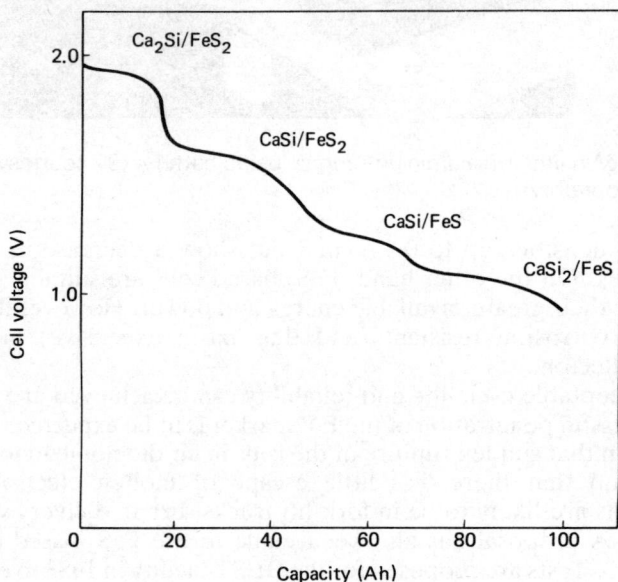

6.15   Discharge behaviour of a Ca$_2$Si/FeS$_2$ cell at 460°C. Electrolyte: LiCl–NaCl–CaCl$_2$–BaCl$_2$.

6.5   High temperature sodium batteries

Sodium melts at 98°C and therefore many of the materials problems experienced with the handling of liquid lithium might be expected to occur in the development of high temperature sodium batteries. However the discovery of solid sodium ion conductors, especially β–alumina, revolutionised the design and fabrication of high temperature cells and has led to a massive research programme into a new type of battery technology over the past fifteen years. The main interest has been in sodium–sulphur batteries and groups in Europe (Chloride Silent Power, British Rail and the

UKAEA in the UK, SAFT and CGE in France, Brown Boveri Corporation in Germany), USA (General Electric Co., Ford Motor Co. and Dow Chemical Co.) and Japan (Toshiba Electric and Yuasa Battery Co.) as well as in Bulgaria, the USSR and the People's Republic of China have been working on the development of this system for EV traction and load levelling applications.

Sodium–sulphur batteries with ß–alumina electrolyte ('Beta batteries')

As shown in Fig. 6.16, the cell is formed in principle by two liquid electrodes, the sodium negative and the sulphur positive, separated by a tube of sintered polycrystalline β–alumina. Since sulphur is an insulator, the compartment

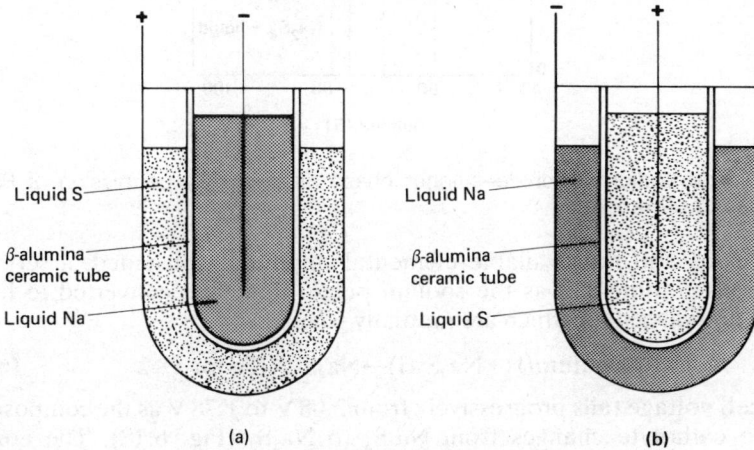

6.16 Schematic diagram of the two basic cell configurations used in beta batteries.

containing the sulphur electrode is fitted with a graphite felt current collector. The cell which may be written as

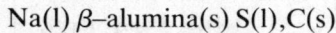

$$Na(l)\ \beta\text{–alumina}(s)\ S(l),C(s)$$

is operated in the temperature range 300–400°C where the reactants and discharge products are in the liquid state and the ionic conductivity of the β–alumina is high. As the cell is discharged, sodium atoms ionise at the surface of the β–alumina tube and migrate through the tube wall to form sodium polysulphides. The initial discharge reaction is

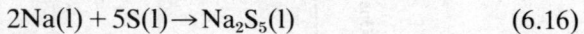

$$2Na(l) + 5S(l) \rightarrow Na_2S_5(l) \tag{6.16}$$

which has an e.m.f. of 2.08 V at 350°C and a theoretical energy density of 790 Wh kg^{-1}. As indicated in the sodium–sulphur phase diagram given in Fig. 6.17, sodium pentasulphide and sulphur are not mutually soluble at the temperature of cell operation, so that two liquid phases are present in the cathode compartment, and the cell voltage is invariant. As the discharge

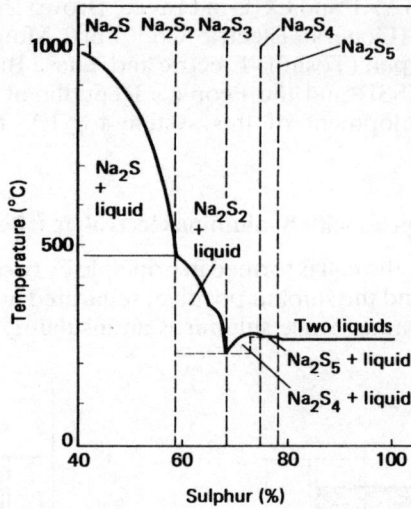

6.17 The sodium sulphide–sulphur phase diagram. (By permission of Power Sources Conference)

progresses and the available elemental sulphur is consumed, a series of reactions commences as the sodium pentasulphide is converted to lower polysulphides, all of which are mutually soluble:-

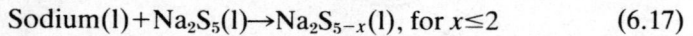

$$\text{Sodium(l)} + \text{Na}_2\text{S}_5(\text{l}) \rightarrow \text{Na}_2\text{S}_{5-x}(\text{l}), \text{ for } x \leq 2 \qquad (6.17)$$

The cell voltage falls progressively from 2.08 V to 1.78 V as the composition of the catholyte changes from Na_2S_5 to Na_2S_3 (Fig. 6.18). The normal working range does not extend beyond Na_2S_3 since if Na_2S_2 and Na_2S were formed, they could crystallise out.

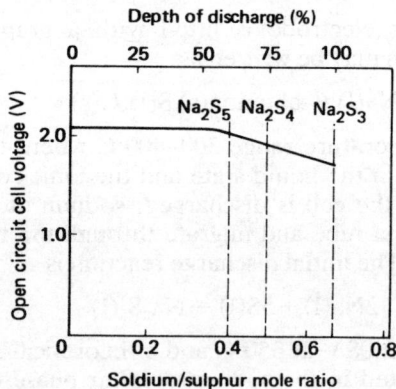

6.18 E.m.f. of sodium–sulphur cell as a function of depth of discharge. (By permission of Power Sources Conference)

As with all high temperature systems, sodium–sulphur cells must be heated up before use. However once they have reached working temperature, iR losses (on discharge or charge) together with $T\Delta S$ contributions during discharge, are sufficient to maintain them at the required temperature. In fact cooling is required under some operating conditions. In EV applications, thermal insulation together with only a very small self-discharge through an internal heater can keep an open circuit sodium–sulphur battery within the operational temperature range for several days.

Cell materials

Almost all practical sodium–sulphur cells are based on electrolytes formed as closed tubes. These are usually manufactured by isostatic pressing, extrusion or slip casting of powdered β–aluminas (or their precursors) followed by batch sintering or continuous zone-sintering at 1600–1700°C. This sintering process strengthens the 'green' tube and increases its density from just over 50% to 99% of the theoretical value. General Electric have developed a successful electrophoretic forming technique to produce a very uniform green tube. The reliability of the electrolyte tube is the key to the success of the sodium–sulphur battery and a great deal of effort is being made to develop dependable units. As can be understood, many of the details of the electrolyte formation process are industrial secrets. One of the most difficult technical problems in the history of this system has been the development of seals to isolate the cell compartments from each other and from the external environment. Generally an insulating α–alumina header is first attached to the β–alumina tube using a glass flux: recently, aluminoborate glass has been shown to be superior for this purpose than the more commonly used silicate glasses. The α–alumina then requires to be bonded to the metal casings: a number of techniques have been employed and a new group of reliable metal-to-ceramic seals have recently been announced. Another materials science problem is concerned with the corrosion of the stainless steel case or the central current collector by molten sodium polysulphides; various coating procedures and materials have been studied.

Cell design and performance

In the most common cell configuration, the sodium is placed inside the electrolyte tube, as in Fig. 6.16(a). This permits the use of smaller diameter tubes and a higher energy density for the cell. In the UK, Chloride Silent Power and British Rail have both chosen a configuration with a central sulphur electrode and a narrow sodium annulus between the electrolyte tube and the cell casing, as in Fig. 6.16(b). For traction applications this latter design has the advantage of significantly increasing the peak power. In addition, the serious problem of corrosion of the steel container by the sodium polysulphide melt is circumvented.

As the discharge proceeds, sodium is removed from the anode chamber, but to ensure a low resistance at the sodium/electrolyte interface it is necessary to keep the entire surface wetted with sodium throughout the discharge. This is achieved in a number of ways, depending on whether the cell is mounted vertically or horizontally. With central sulphur electrode cells the sodium can be kept in an external reservoir and supplied to the annulus surrounding the electrolyte by gravity, gas pressure or by capillary action, using stainless steel mesh wicks (Fig. 6.19(a)–(c)). This last method is particularly successful for horizontally oriented cells, and the sodium electrode resistance is found to remain practically unchanged over the complete range of sodium utilisation. Central sodium electrode cells generally have a gravity feed reservoir above the cell, or store all the sodium required within larger electrolyte tubes containing internal steel wicks (Fig. 6.19(d)–(e)). The design of a practical sodium–sulphur cells is illustrated in

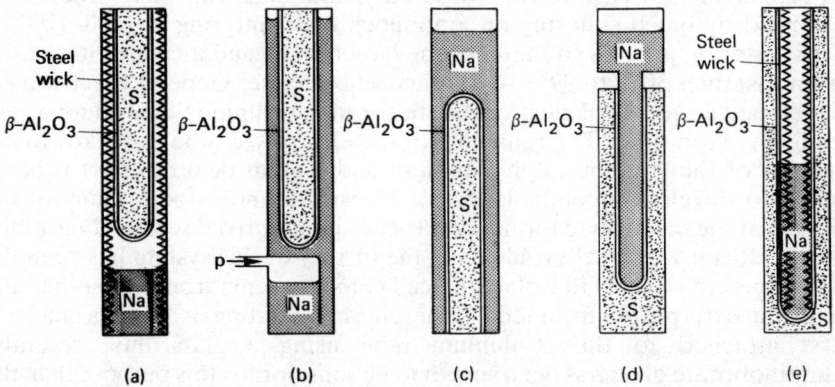

6.19 Schematic diagram of practical beta battery configurations: (a) central sulphur with sodium wick; (b) central sulphur with sodium under gas pressure (in practical cells the gas pressure is produced by incorporating sodium azide pellets – see Fig. 6.20(a); (c) central sulphur with gravity-fed sodium; (d) central sodium with external reservoir; (e) central sodium with internal wick.

Fig. 6.20 and 6.21 and 6.22. Typical charge/discharge curves for Na/S cells with a central sodium electrode after 30 cycles are shown in Fig. 6.23.

For EV applications (e.g. urban delivery vans) the target figures for a commercially viable sodium–sulphur battery would be for a cycle life of 1000 cycles (i.e. 3 years continuous use) and an overall battery practical energy density of $120 \, Wh \, kg^{-1}$, requiring a cell energy density of about $150 \, Wh \, kg^{-1}$. In addition, the design must incorporate adequate safety measures to minimise the effects of a collision or other accident. For load levelling use, requirements are not so strict, but cycle energy efficiency must be high.

Pilot plant manufacture is now in operation so that large scale testing of batteries can be undertaken and a great deal of experience has been gained with laboratory sized cells, full engineering scale cells and commercial prototypes in the 50–200 Ah capacity range. Cell cycle lives of up to 1500 cycles have been reported and cell energy densities of $140 \, Wh \, kg^{-1}$ have

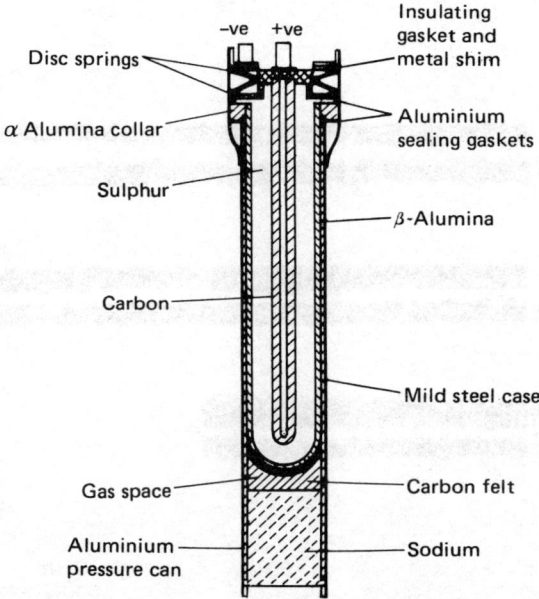

6.20 Cross-section of 38 Ah sodium–sulphur cell (BR16). (By courtesy of British Rail)

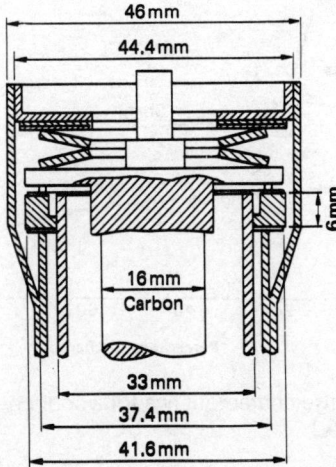

6.21 Detail of compression seal design for 38 Ah sodium–sulphur cell. (By courtesy of British Rail)

been achieved. A number of traction batteries have already been built and tested, with energies of up to 10 kWh, by Chloride Silent Power, British Rail (Fig. 6.24), Brown Boveri, etc. The 7½ tonne van used for road testing sodium–sulphur batteries by Chloride Silent Power is shown in Fig. 6.25.

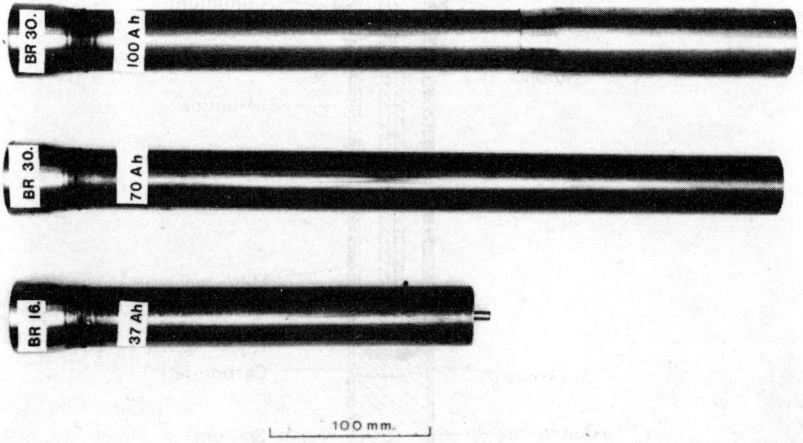

100 mm.

6.22 Practical sodium–sulphur cells. (By courtesy of British Railways Board)

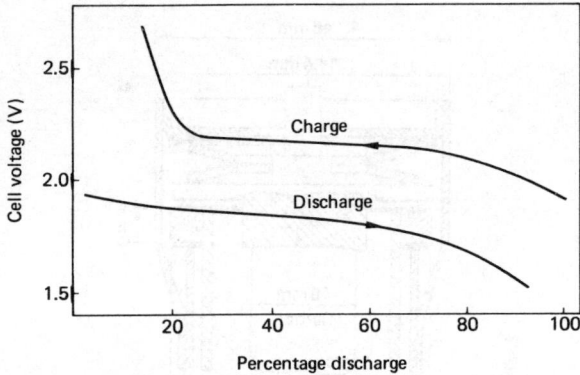

6.23 Charge-discharge characteristics for a sodium–sulphur cell. (By permission of Elsevier Sequoia S.A.)

The final battery configuration for this vehicle will consist of five parallel arrays of 96 cells in series. Road tests using 20 cell prototypes have indicated a range of up to 160 km between charges. General Electric have designed 'modules' containing cells in various series/parallel groupings for load levelling applications. A typical module containing fifty-four cells and having dimensions of approximately 1.30 m × 1.25 m × 0.75 m is illustrated in Fig. 6.26. This company plans to commence extensive load levelling tests

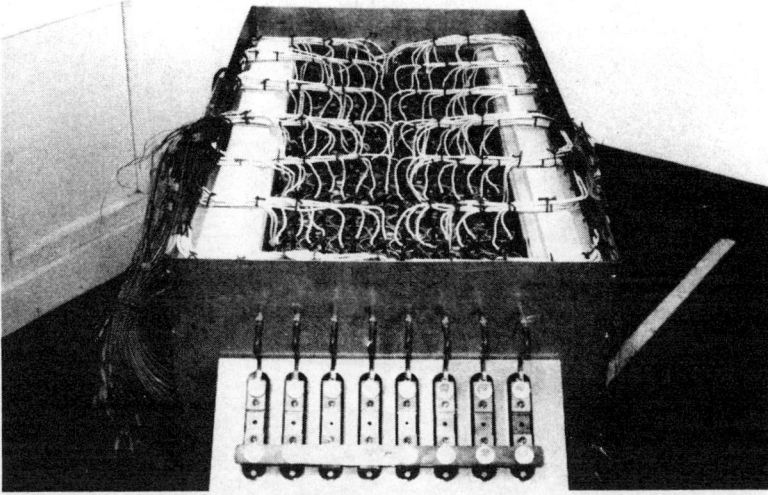

6.24 (a) 10 kWh beta battery consisting of one hundred and eighty-four 38 Ah cells. Leads for monitoring individual cell voltage on left; current shunts at front. (Lid removed for photography)

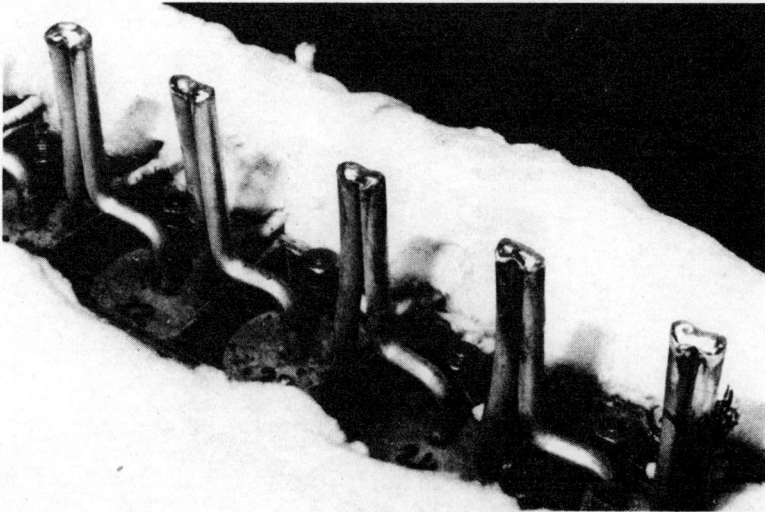

(b) Detail of cell terminals and series connections. (By courtesy of British Railways Board)

in 1983 on a 5 MWh/1 MW battery consisting of forty-eight 360-cell modules. The scale of a 100 MWh/20 MW battery based on 960 modules is shown in Fig. 6.27. In common with many developers. General Electric have adopted the term **'Beta Battery'** to describe this type of unit.

6.25 Purpose built vehicle designed by Chloride Technical Ltd., currently in use for evaluating the road performance of beta battery cells and modules. (By courtesy of Chloride Silent Power Ltd.)

BATTERY
MODULE

GENERAL ⊕ ELECTRIC

6.26 Fifty-four cell beta battery module for load levelling. (By courtesy of General Electric and by permission of 15th Intersociety Energy Conversion Engineering Conference)

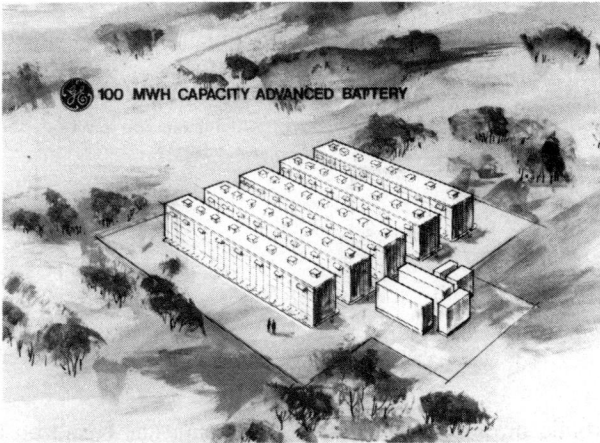

6.27 Artist's impression of a 20 MW/100 MWh beta battery for load levelling applications. (By courtesy of General Electric)

However, despite the considerable advances in the technology of the cell system, some problems still remain. The first concerns the durability of the ceramic electrolyte tube and seals. Studies on the composition and grain microstructure of the β–alumina phase have led to increased reliability, but more work is required before the mechanism of electrolyte failure is completely understood. A second problem is the decline in capacity with cycling, caused mainly by a failure to recharge fully in the two liquid phase region of the sulphur electrode. Capacity retention has been improved by modifying the carbon matrix and current collector, by additives such as SbF_5, Se, etc., to the positive mix and by devising an optimised charging regime.

Sodium–sulphur batteries with glass electrolyte

In the Dow Chemical Co. sodium–sulphur cell, the electrolyte is a sodium conducting borate glass which has been formed into thousands of hollow fibres. The fibres are sealed at one end and held together at the other by a glass header which is connected to a sodium reservoir. The fibres which are up to 10 cm long with an outer diameter of $70\,\mu m$ and a wall thickness of $10\,\mu m$ are filled with sodium and act in parallel to give a sodium/electrolyte interface of very large area. (The normal current density is about $2\,mA\,cm^{-2}$, but cells have been operated successfully at six times this rate.) The sodium filled fibre bundles are immersed in molten sulphur. The cathode current collector is an aluminium foil spiral situated between the rows of fibres (Fig. 6.28), giving a very low cathode–anode spacing. This, together with the low current density, results in negligible concentration polarisation.

Cycle lives exceeding 500 cycles at 80% discharge have been obtained for individual cells. However larger units of 20–40 Ah capacity have much shorter lives, mainly because of failures due to glass breakage. Removal of

Glass fibre tubes

Aluminium foil spiral

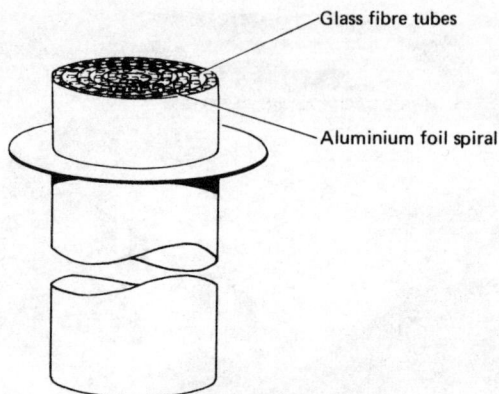

6.28 Schematic diagram of Dow sodium–sulphur cell based on hollow glass fibres filled with sodium.

calcium and sodium oxide impurities from the sodium has proved to be necessary as these were found to be primarily responsible for fibre breakage on deep cycling. A load levelling battery has been designed which will use 12 500 gas-cooled 40 A, 0.8 kWh cells. An EV battery is also under consideration.

Sodium–antimony trichloride, sodium–sulphur (IV) and related batteries

The high operating temperatures of the sodium–sulphur system give rise to severe materials problems connected with corrosion and failure of seals, current collectors, cell housings, etc. In addition, the two phase region of the cathode system may be responsible for some losses in capacity. In the sodium–antimony trichloride battery, a liquid sodium negative electrode and β–alumina electrolyte are combined with $SbCl_3$ dissolved in a low-melting, highly conductive $NaCl$–$AlCl_3$ melt, giving rise to a system with a much lower operating temperature (210–230°C). Despite the lower energy content and higher initial cost, compared with sodium–sulphur, this system is considered a promising alternative for load levelling.

The cell may be written as

$$Na(l)|\beta\text{–alumina}(s)|SbCl_3, NaCl, AlCl_3(l)|C(s)$$

with an overall reaction

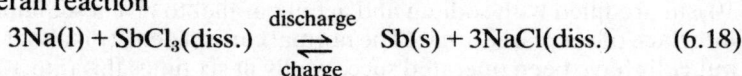

$$3Na(l) + SbCl_3(\text{diss.}) \underset{\text{charge}}{\overset{\text{discharge}}{\rightleftharpoons}} Sb(s) + 3NaCl(\text{diss.}) \qquad (6.18)$$

The e.m.f. is approximately 2.85 V at 210°C and the theoretical energy density is 820 Wh kg^{-1}. The cell is constructed in a similar manner to the sodium–sulphur cell with β–alumina tube and central sulphur electrode. On discharge, sodium ionises at the outer surface of the tube and sodium ions migrate through the wall into the inner (positive) compartment where antimony is discharged. The result of this process is that the melt becomes more basic (i.e. NaCl rich) so that the equilibria of the type:

$$Cl^- + Al_2Cl_7^- \rightleftharpoons 2AlCl_4^- \qquad (6.19)$$

move to the right.

The phase diagram of the $NaCl$–$AlCl_3$ system is shown in Fig. 6.29. The 1:1 mixture (also known as $NaAlCl_4$) melts at about 155°C, and being highly ionised, provides good conductivity in the positive compartment. As NaCl is formed, there is a tendency for it to crystallise out at the temperature of cell operation and this increases cell resistance. Conversely, when the cell is

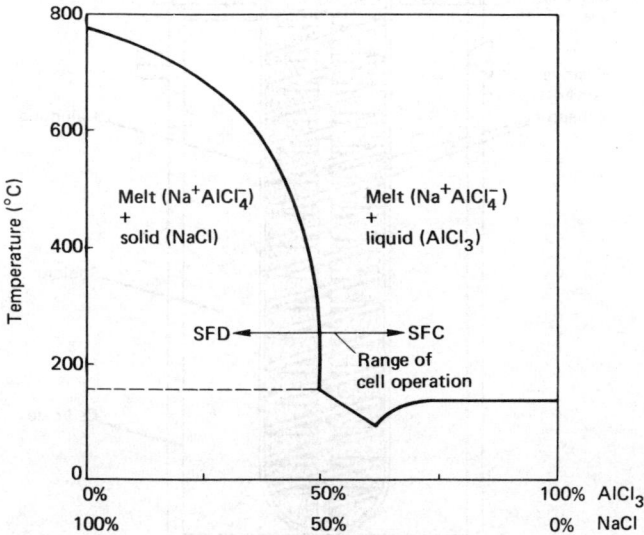

6.29 The $NaCl$–$AlCl_3$ phase diagram in relation to the state of charge of the $Na/SbCl_3$ battery. SFD: state of full discharge; SFC: state of full charge. (By permission of Elsevier Sequoia S.A.)

being charged, an excess of $AlCl_3$ is formed, which again results in higher cell resistance and in addition, since $AlCl_3$ has a significant vapour pressure at 220°C, a pressure build-up in the cell may result. It follows that the performance of the sodium–antimony trichloride cell is greatly influenced by the starting melt composition. Compositions that would ensure more than 52 m/o $AlCl_3$ over the entire discharge range would be desirable in order to avoid sodium chloride crystallisation and the resulting increase in the internal resistance of the cell. This however would imply a large starting excess of $AlCl_3$ with consequent higher resistance and pressure build-up during charging. The most convenient compromise is a composition in which the NaCl content of the melt changes from 35 m/o at full charge to 65 m/o at full discharge. In this condition, the vapour pressure is kept below 1 kPa even at full charge. While some NaCl precipitation is to be expected towards the end of the discharge, the increased resistance protects the cell from over-discharge.

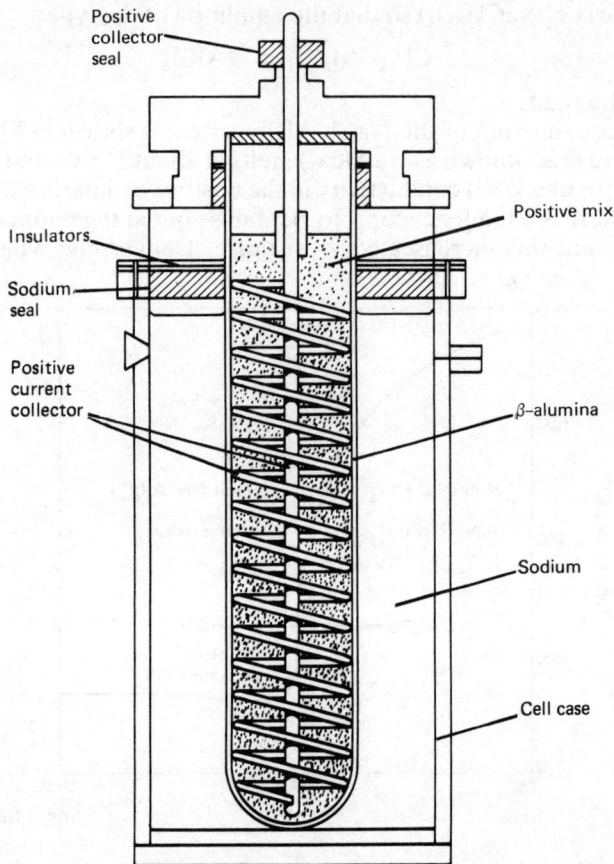

6.30 Cross section of a prototype sodium–antimony trichloride battery.

The prototype of a practical cell assembly is shown in Fig. 6.30. The positive mix contains powdered carbon, with a salt/carbon ratio at full charge of about 10:1 and a spiral tungsten or molybdenum-coated wire, or some form of carbon brush, as current collector. It is found that the design and composition of the latter can have an important effect on the antimony utilisation. The carbon acts as a dispersed current collector and maintains a low internal resistance in the positive electrode system. Because of the lower operating temperature, the problem of seals is not so severe as for sodium–sulphur cells, and it has been found possible to use Teflon seals on the positive side of the cell and silicone rubber on the negative side. However it would appear that glass, ceramic or aluminium gasket seals have better long term durability.

ESB Corporation have studied laboratory cells and 75 Wh engineering prototypes which have the same diameter but about half the length of their proposed 200 Wh load levelling units. Cycle lives of around 500 cycles have been achieved for individual cells, but problems connected with electrolyte cracking and current collector corrosion have been reported. In the final load levelling module, it is planned to immerse 284 tubes in a common sodium pool. Based on the performance of laboratory cells, a practical energy density of $100\,Wh\,kg^{-1}$ is predicted, but has not yet been reported. Charge/discharge curves for a 6 Ah laboratory cell are shown in Fig. 6.31.

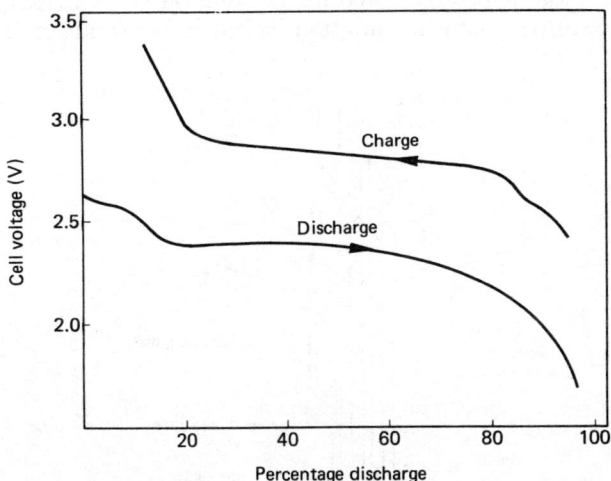

6.31 Charge–discharge characteristics for a 6 Ah sodium–antimony trichloride battery. (By permission of Elsevier Sequoia S.A.)

A number of alternative lower temperature systems, similar to the sodium–antimony trichloride cell, but with improved energy density, have been proposed. In a promising example, which is now being studied, the $SbCl_3$ is replaced by sulphur in an oxidation state of $+4$ and dissolved as SCl_3^+ in the chloroaluminate melt. This cell may be written as

$$Na(l)|\beta-alumina(s)|S(IV),NaCl,AlCl_3(l)|W(s)$$

Here the melt must contain an excess of $AlCl_3$ since oxidation of sulphur to S(IV) is impossible in basic melts. The main overall cell reaction can be written as:

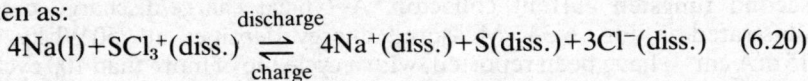

$$4Na(l)+SCl_3^+(diss.) \underset{charge}{\overset{discharge}{\rightleftharpoons}} 4Na^+(diss.)+S(diss.)+3Cl^-(diss.) \quad (6.20)$$

which has an e.m.f. of 4.17 V at 200°C. The main discharge process may be followed by a second process leading to the formation of sulphide ion:

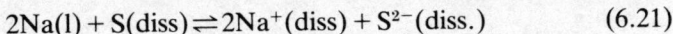

$$2Na(l) + S(diss) \rightleftharpoons 2Na^+(diss) + S^{2-}(diss.) \quad (6.21)$$

As in the case of the Na/SbCl$_3$ system, the positive compartment becomes progressively richer in NaCl as the discharge proceeds, thus requiring the amount of S(IV) present at full charge to be carefully adjusted to match the initial solvent composition. The most favourable situation occurs when the cell is operated so that the four electron process (S(IV)$\rightarrow$S) ends when the solvent composition reaches a NaCl/AlCl$_3$ ratio of 1:1, so that the two electron process (S → S^{2-}) can then proceed in basic (NaCl rich) solvent.

The theoretical energy density of the system ranges from 418 to 306 Wh kg^{-1} as the starting electrolyte composition is varied from 70:30 to 63:37 mole ratio of AlCl$_3$ to NaCl.

In the laboratory cell shown in Fig. 6.32, a stabilised β''–alumina tube contains the molten positive mixture (a solution of SCl$_3$AlCl$_4$ in NaCl–AlCl$_3$, 37–63 m/o), with a tungsten spiral as current collector. The

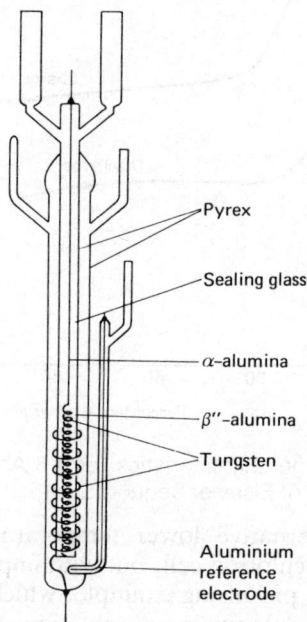

6.32 A sodium–sulphur (IV) laboratory cell. (By permission of the Electrochemical Society)

electrolyte tube is held in a glass envelope containing liquid sodium and a second tungsten current collector. A typical charge/discharge cycle is illustrated in Fig. 6.33. Measured energy densities at 250 Wh kg^{-1} (at 15 mA cm^{-2}) have been reported, with a cycle life of more than 100 cycles.

Studies have also commenced on Se(IV) and a number of metal chlorides in chloraluminate melts. 'Mixed' systems containing both SbCl$_3$ and sulphur in the positive are also being investigated.

6.6 Thermal batteries

A thermal battery is a device which is able to supply energy only after it has been activated by the rapid application of heat. This condition can be achieved with systems based on electrolytes which change in a discontinuous manner from a low to a high conductivity state when heated. Such an

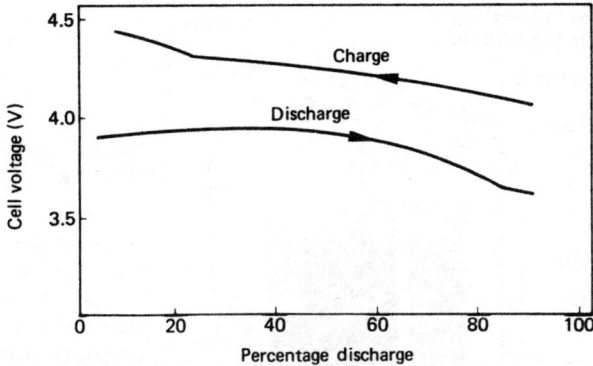

6.33 Charge/discharge characteristics for a sodium–sulphur (IV) cell at 220 °C. Charge: 8.6 mA cm⁻²; Discharge: 17.2 mA cm⁻². (By permission of the Electrochemical Society)

electrolyte may typically be a solid with a phase transition to an ionically conductive phase (considered in Chapter 7) or a salt mixture with a low melting point. For example, the LiCl–KCl eutectic mixture described above is solid and chemically and electrochemically inert at room temperature, but becomes highly conductive on melting at 352°C. A thermal battery is thus a **primary reserve system**, which in its inert condition has a practically unlimited storage life, is usually constructed to be resistant to mechanical stress, humid environments etc., but which is capable of supplying a single high power discharge whenever required. The discharge pulse may be designed to last from a few tens of seconds up to an hour. Thermal activation is generally obtained using in-built pyrotechnic heat sources ignited by an electrical signal, and the electrolyte can be melted in a few tenths of a second (1–3 s for larger batteries). Thermal batteries are used mainly for military, aerospace and emergency applications.

A practical lithium–iron disulphide thermal battery developed by Sandia National Laboratories is shown in Fig. 6.34. This typical reserve battery uses the LiCl–KCl eutectic as electrolyte, LiSi alloy as anode material and FeS_2 as cathode. (The electrochemical characteristics of these components were described above.) The anode is prepared by directly pressing finely powdered lithium–silicon alloy into a cylindrical pellet. The electrolyte pellet is formed similarly using a mixture of LiCl–KCl eutectic and MgO binder (65%–35%) – the latter immobilises the electrolyte when molten and acts as a separator. The cathode pellet is a mixture of 64% FeS_2, 16% LiCl–KCl eutectic and 20% electrolyte binder (here SiO_2). The heat pellet associated with each cell is a mixture of 88% iron powder and 12% potassium perchlorate. This pellet is slightly larger in diameter than the others, to improve contact with the fuse ignition strip (formed by a slurry of $Zr/BaCrO_4$ mixture). The anode and cathode current collectors are stainless steel discs, which also act as thermal buffers and prevent anode and cathode materials from reacting with the heat pellets.

Cells are assembled in the following sequence: anode collector, anode pellet, electrolyte pellet, cathode pellet, cathode collector and heat pellet.

Match support
(fibre frax laminate)
Electric match
Zr/BaCrO$_4$ (Pad)
Negative
Mica
Thermal insulation
(Min-K TE1400)
Positive
Mica
Fibre-frax wrap
Anode collector
Anode (Li(Mx))
Electrolyte binder
Cathode (FeS$_2$)
Cathode collector
Heat pellet (Fe/KClO$_4$)
Mica
Zr/BaCrO$_4$ (Fuse Strip)
Fibre frax (Pad)

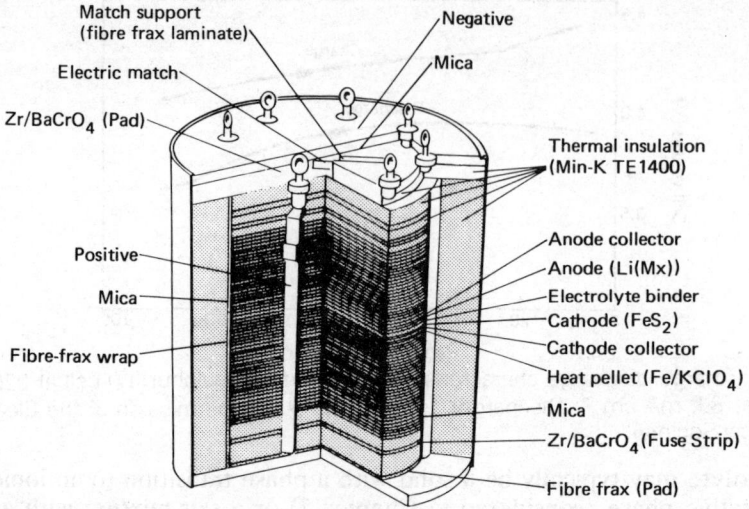

6.34 Cross-section of 0.5 A, 28 V lithium–iron disulphide 60 minute thermal battery. The battery contains 15 active cells and has a total volume of 400 cm³. (By courtesy of Sandia National Laboratories)

The battery is formed by pressing together fifteen of these single cells. The assembly is housed in a closed cylindrical container provided with good thermal insulation and glass-metal sealed top feedthroughs for the external leads and for the electric firing signal which ignites the fuse strip.

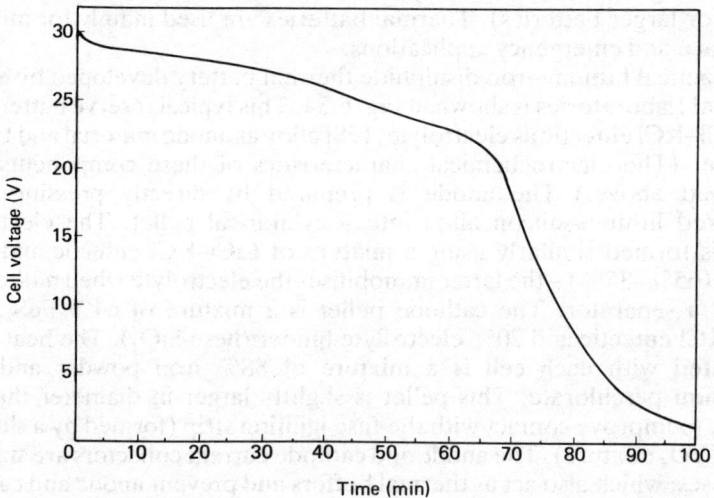

Cell voltage (V)

Time (min)

6.35 Constant load discharge of lithium–iron disulphide thermal battery. Load resistance 56 Ω. (By courtesy of Sandia National Laboratories)

When activated, the battery supplies the typical discharge current shown in the discharge curve in Fig. 6.35. The beginning of the discharge in this, and more generally in all thermal batteries, is characterised by a peak voltage which falls off as the internal resistance of the cell rises, as a result of a build-up of insoluble products at the electrodes and a fall in temperature of the electrolyte. Usually the life of a thermal battery is measured as the time elapsed to reach a fixed fraction (e.g. 80%) of the peak voltage.

Other systems

Other common anode materials for thermal batteries are lithium alloys, such as Li/Al and Li/B, lithium metal in a porous nickel or iron matrix, magnesium and calcium. Alternative cathode constituents include $CaCrO_4$ and the oxides of copper, iron or vanadium. Other electrolytes used are KBr–LiBr and, more generally, all lithium halide systems which are used particularly to prevent electrolyte composition changes and 'freezing out' at high rates when lithium-based anodes are employed.

The calcium–calcium chromate thermal cell has been established for many years. In the LiCl–KCl eutectic, the reaction product of this cell is a mixed lithium–calcium–chromium oxide. However this system cannot provide as high specific capacity or energy density as the lithium-based systems described above. Furthermore, it suffers from parasitic chemical reactions which are exothermic and often uncontrolled.

7 Solid state cells

7.1 Introduction

The concept of an all solid-state battery is very appealing since such a system would possess a number of very desirable characteristics – e.g. absence of any possible liquid leakage or gassing, the likelihood of extremely long shelf-life and the possibility of operation over a wide temperature range. Solid-state batteries could be constructed with excellent packaging efficiency for the active components, without separators and using simple lightweight containers. The opportunities for extreme miniaturisation and very simple fabrication techniques are of obvious importance in applications where size and reliability are key factors, as for example in implantable electronic instrumentation such as cardiac pacemakers, physiological monitoring/telemetry packages, etc. However to be of practical value, a solid-state cell must also fulfil, to some extent, the basic requirements of all effective power sources, namely sufficiently low internal resistance and high voltage to give adequate levels of power, together with an acceptable cost. The first of these requirements is primarily a function of the electrolytic phase, which must therefore be a solid material characterised by high ionic conductivity with negligible electronic mobility: the former leads to low internal resistance of the cell, and the latter prevents self-discharge, thus promoting long shelf-life.

A 'solid electrolyte' is a phase which has an electric conductance wholly due to ionic motion within the solid lattice. Such phases have been known for over a century, but until recently all known materials of this type had high resistivities at ambient temperatures. This restricted the development of solid-state devices to a limited number of laboratory cells, used for thermodynamic studies, and of little interest as power sources. A striking development occurred towards the end of the 1960's with the discovery of a series of solids of general formula MAg_4I_5 (for $M = Rb,K, \ldots$), having an exceptionally high ionic conductivity ($> 10\,Sm^{-1}$ at room temperature). For a time these were referred to, rather inappropriately, as 'superionic conductors'.

A number of structural features have been found to characterise solids with high ionic conductivity and to distinguish them from the more usual ionic crystals. Ionic transport in 'normal' salt lattices at high temperatures takes place by a defect mechanism, and the relatively poor conductivity of such phases is largely due to the small equilibrium concentration of defects present. Generally the structures of solid electrolytes are not close-packed, but contain two- or three-dimensional networks of passageways interlaced between interconnected polyhedra of the fixed ions, and through which selected mobile ions may move. The number of sites available for the mobile ions is much larger than the number of mobile ions themselves, and the solid has therefore a characteristic highly disordered structure. In $RbAg_4I_5$, for example, the sixteen silver ions in each unit cell are distributed over fifty-six available sites.

The high conductivity is due to a combination of

(i)   a high concentration of mobile ions, and
(ii) a low activation energy for ionic motions from site to site.

One of the main advances in understanding solid electrolytes has stemmed from crystallographic studies which identified the types and numbers of sites available to the diffusing ions and emphasised the significance of the spatial relationship between sites. Mobility is enhanced when the mobile ion sites form a network of channels or a plane through which the ions can move, and the 'simplicity' of these passageways has an important influence on the value of the conductivity. In the crystalline solid electrolyte $Ag_{26}I_{18}W_4O_{16}$, the conduction pathways involve ninety face-sharing iodide polyhedra and fifty-six mixed iodide–oxide polyhedra per unit cell. Some silver ion pathways in this structure are shown in Fig. 7.1. The mobile ions are considered to move between low energy, highly coordinated sites by passing through a shared

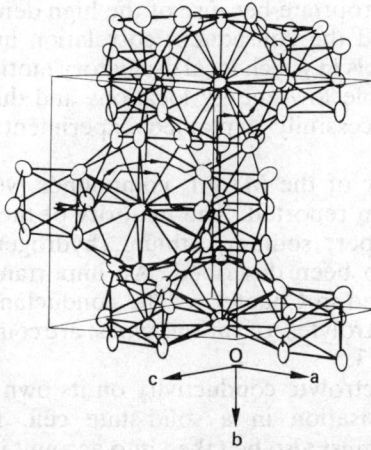

7.1   The iodide arrangement in crystalline $Ag_{26}I_{18}W_4O_{16}$: some pathways for silver ion motion are shown by the arrows. (By permission of Academic Press)

polyhedral face of only slightly higher energy. In Fig. 7.2, a number of inter-linked tetrahedra formed by fixed ions is shown. A mobile ion traversing this structure in a vertical direction would pass through trigonal co-ordination on its way between neighbouring tetrahedral sites.

The second necessary condition for a solid to have high ionic conductivity is that the mobile ions have a high diffusion coefficient i.e. it is indeed a 'fast

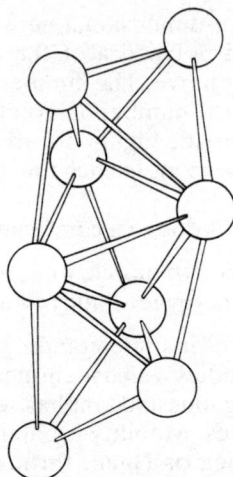

7.2 Interconnected iodide ion tetrahedra providing a passageway for ion migration: mobile silver ions must pass through a three-coordinated configuration on their way between tetrahedral sites.

ion conductor'. Much attention has been given to developing models of ionic motion. The simple hopping models applied successfully in the case of defect transport are not appropriate because of the high density of mobile ions in solid electrolytes, and the consequent correlation in the motion. Quasi-liquid sublattice models in which local oscillatory motion takes place during residence in favourable low energy locations and during transit between such sites, have successfully explained experimental observations in a number of cases.

Since the discovery of the MAg_4I_5 compounds, very many more solid electrolytes have been reported. The majority of these are also silver ion conductors, but copper, sodium, lithium, hydrogen and fluoride solid electrolytes have also been described. An important example is sodium β–alumina as discussed in Chapter 6. The conductance values of selected examples of solid electrolytes, of various types, are compared as functions of temperature in Fig. 7.3.

However, high electrolyte conductivity on its own does not necessarily guarantee low polarisation in a solid-state cell. Electrode/electrolyte interfacial resistance must also be taken into account in this respect, and in contrast to the more familiar situation with conventional aqueous systems where the solid electrodes are uniformly wetted by the liquid electrolyte, the all solid configuration of the cell may create non-uniform contact at the interfaces. Differential expansion and contraction of electrodes and electrolyte may lead to poor contact (and consequent high internal resistance due to low effective electrode/electrolyte interfacial area) or even to a complete open circuit during cell operation. The situation is even more serious with secondary cells, as illustrated schematically in Fig. 7.4, where the effects of non-uniform discharge, and the deposition of metal in

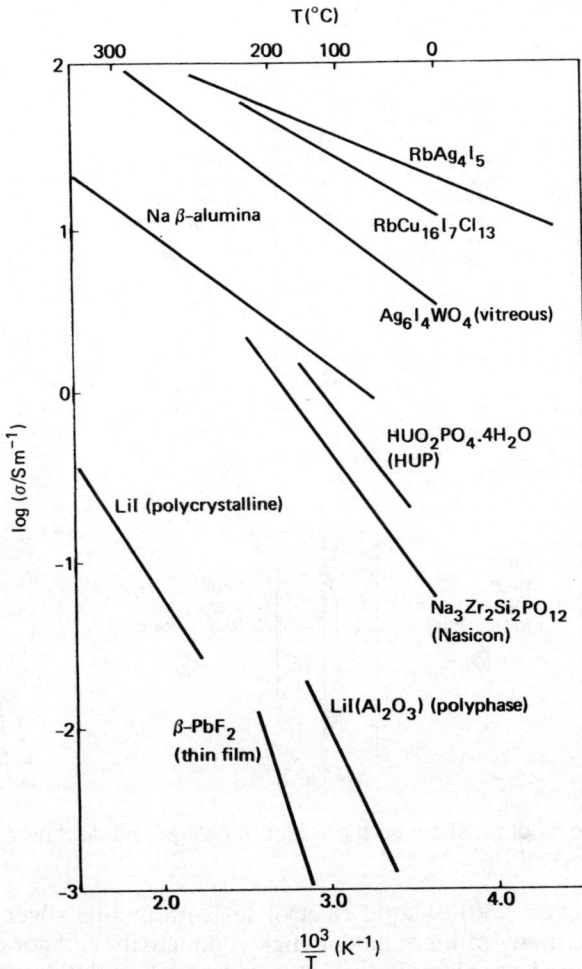

7.3 The conductance of some typical solid electrolytes as a function of temperature.

dendritic form during charge are shown. Indeed interfacial polarisation phenomena are among the most severe problems in the development of practical solid-state cells. A common expedient is to mix the electrode reactant with the electrolyte in order to form an electrodic mass (e.g. as a compressed powder) with a greatly enlarged interfacial area. This reduces the current density at the interface during operation of the cell and thus alleviates the polarisation. Recently secondary solid-state cells have been proposed which use 'insertion electrodes', where the mobile ionic species may be accepted by or expelled from an electronically conducting host lattice which retains its major structural features throughout the operating cycle. However such cells are still at an early stage of development, and most practical ambient solid-state cells are of the primary type.

Discharge :- $M \longrightarrow M^{n+} + ne$

Metal (M)

Electrolyte

Initial stage (loss of contact)

Later stage (increased loss of contact)

Charge :- $M^{n+} + ne \longrightarrow M$

Initial stage (dendrite formation)

Later stage (cell fracture)

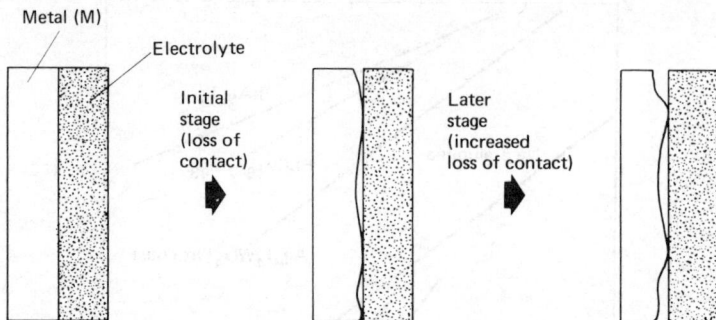

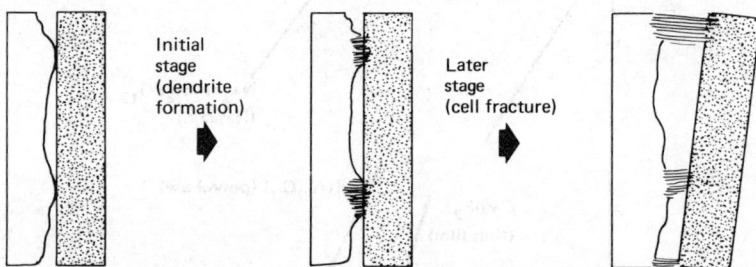

7.4 Schematic diagram showing the effect of charge and discharge on a metal (M)|solid electrolyte (E) interface.

A wide variety (> 50) of solid electrolytes with mobile silver ions have been reported, many of them having high conductivity and good stability. The majority of these electrolytes are based on silver iodide and may be regarded as 'cation modified', e.g. $RbAg_4I_5$ or 'anion modified', e.g. $Ag_7I_4AsO_4$. The decomposition potential of these materials (i.e. the voltage difference which produces significant electrolysis) is found to be very close to that of silver iodide itself, viz. $0.69\,V$ at $25°C$. This restricts the usable e.m.f. of single silver solid-state cells to a value which is generally too low to meet the requirements of most electronic instrumentation. Commercial exploitation of silver-based cells is also inhibited by the high cost of silver, so that their practical utilisation is limited to very specific applications.

Cells based on the transport of ions which have higher energy and lower cost, have also been studied. A number of copper solid electrolytes have been characterised, but their stability has generally been found to be poor, especially in the presence of air or moisture. More attention has been paid to the search for lithium solid electrolytes because of the possibility of developing a high voltage solid-state cell which might have a performance comparable with the lithium–organic cells described in Chapter 5. (As pointed out above, in certain applications the use of all solid cells is desirable because of their inherently greater safety and reliability.) However despite reports on a wide range of lithium solid electrolytes, no really good ambient lithium ion conductor has yet been discovered.

In its most general configuration, a solid-state cell may be formed by finely grinding the constituents and then sequentially pressing the anode-electrolyte mixture, the electrolyte and the cathode-electrolyte mixture (with or without 5–10% of a binding agent) in a suitable die to form a single composite pellet. When the electrolyte is of the ceramic type this procedure is not feasible and a sintering process has to be used instead. Since the electrolyte does not participate in the cell reaction it is sometimes possible to reduce its thickness to that of a thin film. Under these circumstances it is possible to replace a highly conductive solid electrolyte with, say, a defect ionic conductor of much higher resistivity.

The only commercial ambient solid-state batteries so far produced have been based on either silver or lithium anodes and these will now be described. Various other systems have been studied, and although the majority of them are only of academic interest at present, some of these will be briefly outlined.

7.2 Silver–iodine batteries

The first commercial solid-state battery was manufactured at the end of the 1960's in the USA by Gould Ionics: this was a silver–iodine battery using $RbAg_4I_5$ as electrolyte. An essential constraint on any cell system is that the active components must not react with the electrolyte either directly or by electrolytic action. Free elemental iodine reacts with $RbAg_4I_5$, degrading it to poorly conducting phases by the process

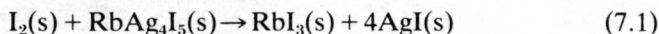

$$I_2(s) + RbAg_4I_5(s) \rightarrow RbI_3(s) + 4AgI(s) \tag{7.1}$$

Thus the cell

$$Ag(s)|RbAg_4I_5(s)|I_2(s),C(s)$$

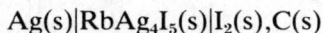

cannot be formed in an equilibrium condition. Further, since the e.m.f. of such a cell at 25°C would be 0.687 V, which is higher than the decomposition potential of $RbAg_4I_5$ (0.660 V), electrolytic process such as

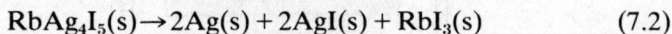

$$RbAg_4I_5(s) \rightarrow 2Ag(s) + 2AgI(s) + RbI_3(s) \tag{7.2}$$

would occur in the cell. The iodine must therefore be complexed to reduce its activity. In the Gould system, rubidium triiodide was used as the iodine source, to give the cell

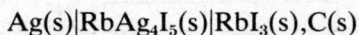

$$Ag(s)|RbAg_4I_5(s)|RbI_3(s),C(s)$$

which has an e.m.f. of 0.660 V at 25°C and a theoretical energy density of 48 Wh kg^{-1}. Since $RbAg_4I_5$ is thermodynamically unstable at temperatures below 27°C, two overall cell reactions have been proposed:

Above 27°C:

$$14Ag(s) + 7RbI_3(s) \rightarrow 3RbAg_4I_5(s) + 2Rb_2AgI_3(s) \qquad (7.3.)$$

and below 27°C:

$$4Ag(s) + 2RbI_3(s) \rightarrow Rb_2AgI_3(s) + 3AgI(s) \qquad (7.4)$$

The original Gould silver–iodine battery had a very limited range of application because of its low voltage and high cost, so that its production has now ceased. However it is of great historical importance as it greatly stimulated further development in the area of all solid-state systems. A typical configuration of this battery is shown in Fig. 7.5. The anode is a blended mixture of powdered silver, carbon and electrolyte. High

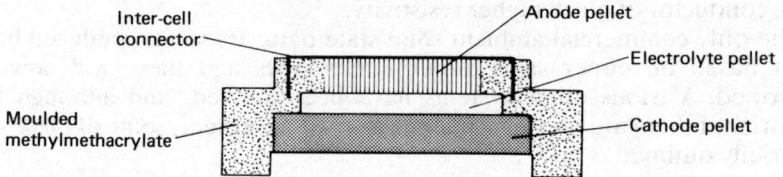

7.5 Ag–RbI₃ solid state cell. (By permission of John Wiley and Sons)

efficiency, coupled with low polarisation is obtained by forming the electrode by *in situ* reduction of Ag_2O by carbon. The cathode is a blended mixture of RbI_3, $RbAg_4I_5$ and carbon. The electrolyte layer may contain up to 10% of a polycarbonate resin as binding agent. The cell components are sequentially pressed into a single three layer pellet. A series of pellets can be combined to form a higher voltage battery, as shown in Fig. 7.6. Because the

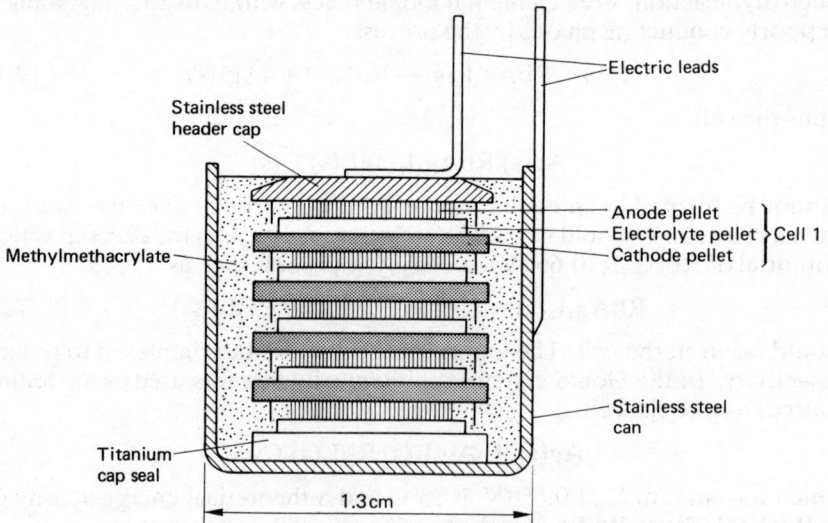

7.6 Ag–RbI₃ solid state battery. (By permission of John Wiley and Sons)

electrolyte disproportionation reaction:

$$2RbAg_4I_5(s) \rightarrow 7AgI(s) + Rb_2AgI_3(s) \qquad (7.5)$$

is catalysed by water vapour, the cell components must be handled in a dry atmosphere and the battery casing must be hermetically sealed.

Typical discharge curves of the silver–iodine solid-state battery, shown in Fig. 7.7, have flat plateaus even at moderately high current drains, and good cathodic utilisation ($\approx 85\%$).

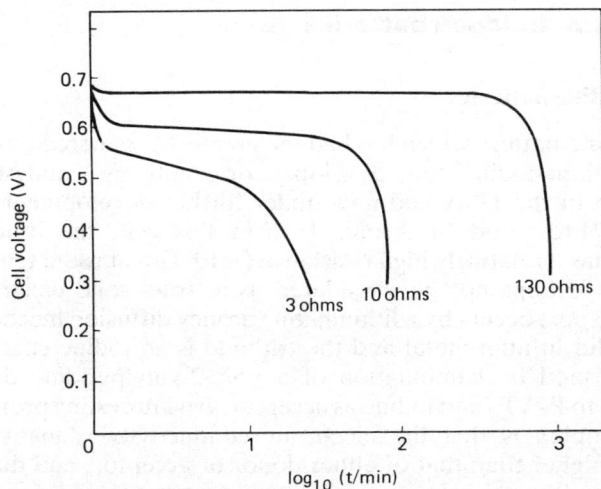

7.7 Discharge characteristics for Ag–RbI$_3$ solid state cells under various loads.

For a cell with a rated capacity of 25 mAh, a practical energy density of 4.4 Wh kg^{-1} (15 Wh dm^{-3}) was obtained at a current drain corresponding to 25 μA cm^{-2}. The cell reaction implies the formation of poorly conducting Rb$_2$AgI$_3$ and AgI at normal cell operating temperatures, and this would suggest an increasing iR drop and cathode polarisation (with consequent loss of energy) as the discharge progresses. In practice, the cathode products are not formed as a compact layer, but are dispersed within the RbAg$_4$I$_5$ and carbon conducting phases which constitute the composite electrode.

Losses in capacity during storage have been identified as being due mainly to the diffusion of iodine through the electrolyte to the anode, but no significant degradation of capacity or increase in cell resistance is experienced with sealed cells stored at $< 23°C$ over a period of more than ten years. Experiments have been carried out in which an 'iodine getter' composed of finely divided silver powder and Rb$_2$AgI$_3$ was incorporated in the electrolyte layer so that diffusing iodine reacted according to

$$7I_2(\text{diffusing}) + 14Ag(s) + 2Rb_2AgI_3(s) \rightarrow 4RbAg_4I_5(s) \qquad (7.6)$$

The modest energy values and high cost of silver–iodine batteries have prevented further commercial exploitation of this system. However some interesting studies have been made in which alternative cathode materials were examined. The use of $(CH_3)_4NI_9$ and $(CH_3)_4NI_5$ provided cathodes with a higher weight-per-cent of available iodine, together with a suitable iodine activity. In addition one of the discharge products, $((CH_3)_4N)_2Ag_{13}I_{15}$ is a silver ion conductor, which contributes to the maintenance of low polarisation. Another suggestion has been the use of electronically conducting iodine charge transfer complexes such as perylene–iodine or phenothiazine–iodine, which do not require carbon additives.

7.3 Lithium–halogen batteries

Lithium–iodine batteries

The solid-state battery which has had the greatest commercial success is the primary lithium–iodine cell, developed originally by Catalyst Research Corporation in the USA and now under further development by Wilson Greatbatch Ltd., and Medtronic Inc. In this cell, the lithium iodide electrolyte has a relatively high resistance ($\approx 10^5$ Ωm at room temperature), and should perhaps not be considered as a 'true' solid electrolyte since electric transport occurs by a lithium ion vacancy diffusion mechanism. The anode is solid lithium metal and the cathode is an iodine charge transfer complex formed by combination of a poly-2-vinylpyridine donor (here abbreviated to P2VP) and iodine as acceptor. An interesting property of this type of complex is that the electronic conductivity is many orders of magnitude higher than that of either donor or acceptor, and that this high conductivity is maintained over a wide range of donor/acceptor compositions. This behaviour is shown in Fig. 7.8 for the $P2VP.nI_2$ complex. The complex is easily formed by direct reaction of iodine with P2VP powder to form a highly viscous tar which can be poured when hot, or a semi-solid mass. It generally contains over 90% by weight of iodine and has an iodine vapour pressure close to that of elemental iodine, so that it makes a very convenient iodine electrode.

Lithium–iodine cells are produced simply by making direct contact between anode and cathode. On touching, a thin layer of lithium iodide is formed by direct reaction. As soon as the layer becomes complete, the reaction rate decreases sharply since diffusion of iodine through lithium iodide is very slow. Thus the cell may be written as:

$$Li(s)|LiI(s)|P2VP.nI_2(s)$$

That the cell reaction may be written simply as

$$Li(s) + \tfrac{1}{2}I_2(s) \rightarrow LiI(s) \tag{7.7}$$

is supported by its measured e.m.f. of 2.80 V at 25°C, since this value agrees exactly with that calculated for the reaction: Latimer's 'Oxidation Potentials' gives the free energy of formation of LiI(s) at 25°C as $268\,kJ\,mol^{-1}$, corresponding to an e.m.f of 2.79 V. More accurately, the cell reaction might be written as

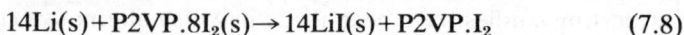

$$14Li(s) + P2VP.8I_2(s) \rightarrow 14LiI(s) + P2VP.I_2 \tag{7.8}$$

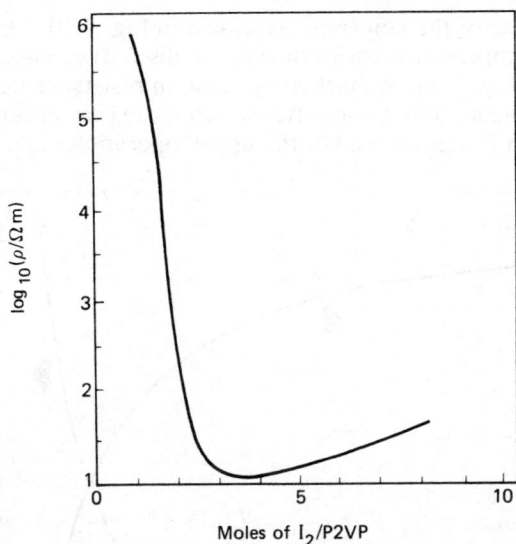

7.8 Resistivity of the poly–2–vinylpyridine iodine charge transfer complex as a function of iodine content. (By courtesy of Catalyst Research Corporation)

On discharge, lithium ionises at the anode/electrolyte interface and lithium ion transport through the lithium iodide occurs by the motion of a lithium ion vacancy in the direction from cathode to anode, as shown schematically in Fig. 7.9. As the discharge progresses, further LiI forms at the cathodic side of the electrolyte, and as the LiI layer thickens, so the

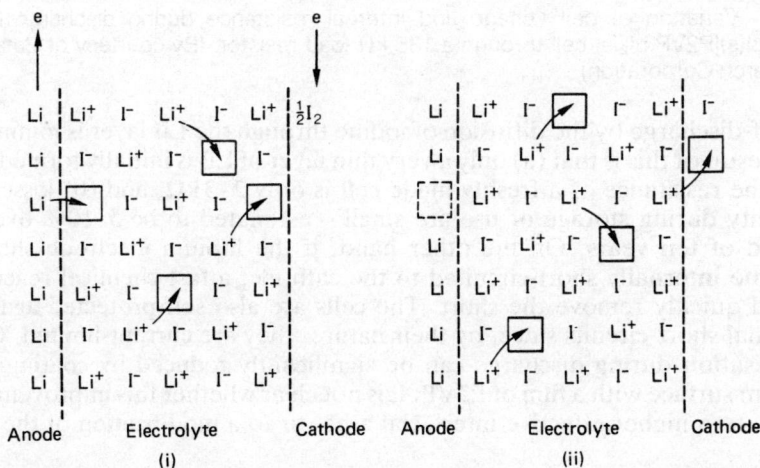

7.9 Mechanism of lithium ion transport by vacancy motion in lithium iodide. Cell discharge is accompanied by the migration of lithium ion vacancies in the direction from cathode to anode.

internal resistance of the cell rises, as shown in Fig. 7.10. The resistivity of the P2VP.nI$_2$ complex also varies during the discharge, since it depends on iodine content (Fig. 7.8). A marked increase in resistance occurs when the I$_2$/P2VP ratio reaches about two, after which there is an abrupt decay in cell voltage, which indicates the end of the useful operating range of the cell.

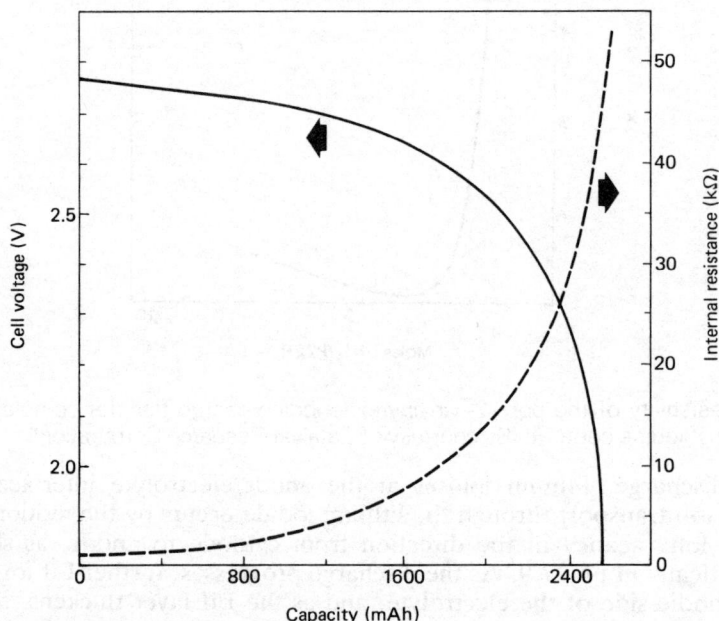

7.10 Variation of cell voltage and internal resistance during discharge of a Li(s)|LiI(s)|P2VP.nI$_2$(s) cell through a 135 kΩ load resistor. (By courtesy of Catalyst Research Corporation)

Self-discharge by the diffusion of iodine through the LiI layer is minimal. The result of this is that (a) only a very thin layer of LiI is initially formed, so that the resistance of a freshly made cell is only 2–3 kΩ, and (b) losses in capacity during storage or use are small – estimated to be 5–10% over a period of ten years. On the other hand, if the lithium electrode should become internally short-circuited to the cathode, a fast chemical reaction would quickly remove the short. The cells are also self protected against external short-circuits since, by their nature, they are current-limited. Cell polarisation during discharge can be significantly reduced by coating the lithium surface with a film of P2VP: it is not clear whether this improvement is due to a higher effective interfacial area, or to a modification of the LiI layer.

Lithium–iodine cells are assembled in a controlled environment and are hermetically sealed with welded cases and glass-to-metal or ceramic-to-metal seals. They are manufactured in three basic configurations: as normal button cells, as button cells for direct mounting on printed circuit boards,

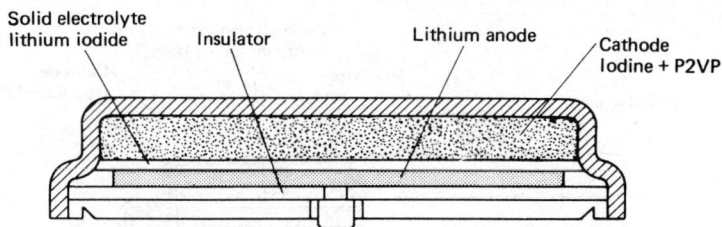

7.11 Lithium–iodine solid state button cell for electric watches and pocket calculators. (By courtesy of Catalyst Research Corporation)

and as cardiac pacemaker batteries. In Fig. 7.11 the cross-section of a button cell, used mainly in electric watches and pocket calculators, is shown. Such cells are produced with rated capacities in the range 120–250 mAh and are designed for current drains of 0–60 μA. Typical discharge curves are shown in Fig. 7.12. At low drains ($\approx$1 μA cm^{-2}), a practical energy density of 120 Wh kg^{-1} (600 Wh dm^{-3}) is obtained. The cells may be operated over a wide temperature range, although the maximum discharge current falls off logarithmically with decreasing temperature.

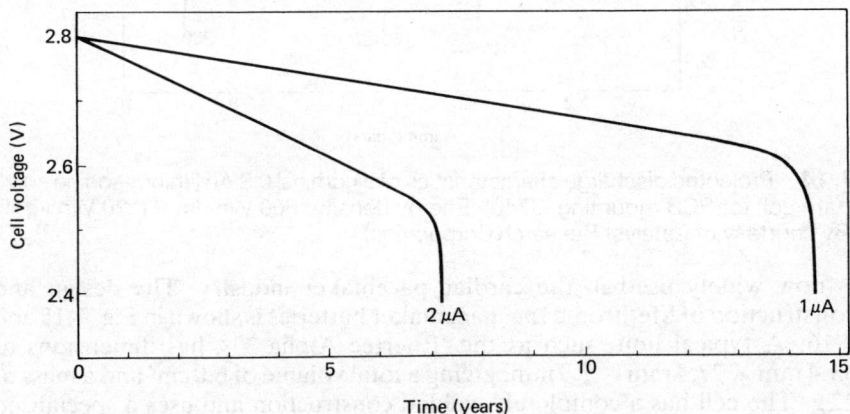

7.12 Projected discharge characteristics of a nominal 120 mAh lithium–iodine solid state button cell (S19P-20). Weight of cell: 2.8 g; volume of cell: 0.56 cm³. (By courtesy of Catalyst Research Corporation)

Larger button cells with rated capacities in the range 460–870 mAh are manufactured for direct mounting on printed circuit boards where they are used as standby power sources for CMOS RAMs, reference voltage sources, etc. Fig. 7.13 shows the construction of these cells and the position of the terminal/mounting pins. Projected discharge curves for a Catalyst Research Corporation 870 mAh cell are given in Fig. 7.14.

The high reliability and the complete absence of faults such as electrolyte leakage or gas generation make the lithium–iodine solid-state battery a particularly suitable device for powering implanted electronic devices, and it

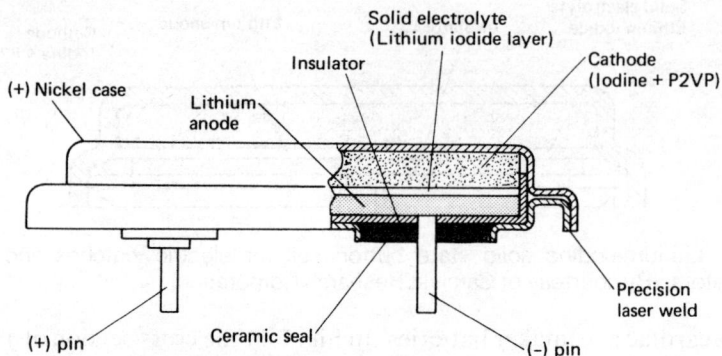

7.13 Construction of a lithium–iodine solid-state cell for PCB mounting. (By courtesy of Catalyst Research Corporation)

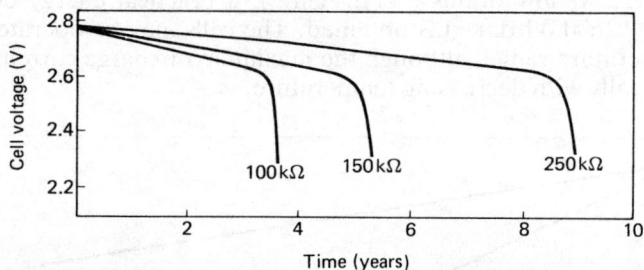

7.14 Projected discharge characteristics of a nominal 0.8 Ah lithium–iodine solid-state cell for PCB mounting (3740). Energy density: 600 Wh dm^{-3} (120 Wh kg^{-1}). (By courtesy of Catalyst Research Corporation)

is now widely used in the cardiac pacemaker industry. The design and construction of Medtronic Inc. pacemaker batteries is shown in Fig. 7.15 and 7.16. A typical unit, such as the 'Enertec Alpha 33', has dimensions of 33.4 mm × 27.4 mm × 7.9 mm, giving a total volume of 6.0 cm^3 and a mass of 22 g. The cell has a completely welded construction and uses a specialised glass-to-metal seal for the electrical feedthrough. The projected capacity as a function of current drain, determined on the basis of a mathematical model, is shown in Fig. 7.17. The upper line represents the theoretical capacity of 2.6 Ah, assuming 100% utilisation of the iodine present in the P2VP.8I$_2$; since each vinyl pyridine entity reacts irreversibly with approximately 0.5 mol of iodine, the maximum available capacity is reduced to 2.1 Ah, giving a maximum energy density of 290 Wh kg^{-1} (960 Wh dm^{-3}). The practical capacity depends on the rate of discharge. At very low rates ($\approx 10 \mu A$) self-discharge processes limit the capacity, while at high rates ($\approx 100 \mu A$) polarisation becomes dominant.

Lithium–bromine batteries

A lithium–bromine charge transfer complex cell was patented in 1976. The

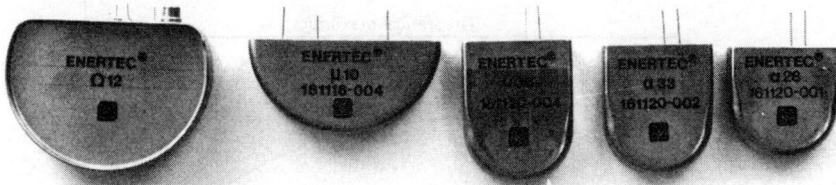

7.15 Range of lithium–iodine solid state pacemaker cells. (By courtesy of Medtronic Inc.)

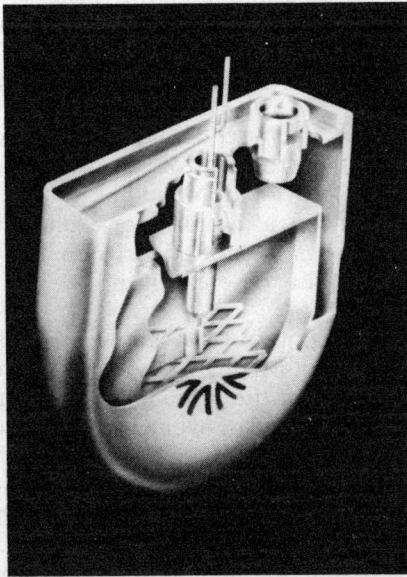

7.16 Cutaway of a lithium–iodine pacemaker cell. (By courtesy of Medtronic Inc.)

solid electrolyte is LiBr, and the cell has an e.m.f. of 3.50 V. Trials by W. Greatbach Ltd., indicate that a practical energy density of 1250 Wh dm^{-3} can be achieved at low rates. Lithium bromide is however fifty times more resistive than lithium iodide, and diffusion of bromine in LiBr is faster than that of iodine in LiI.

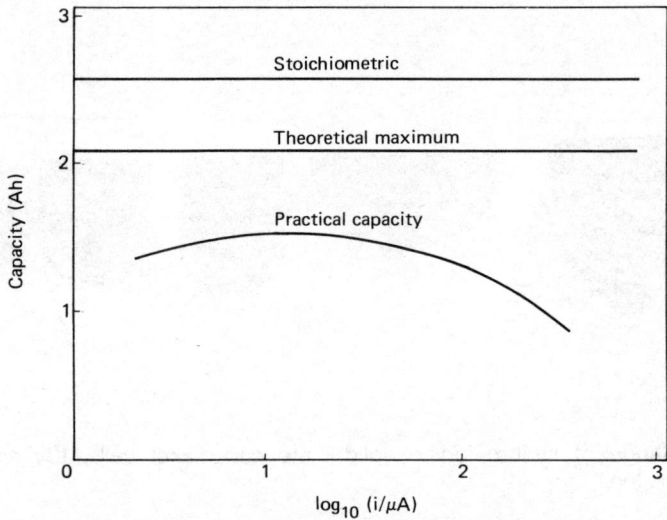

7.17 Projected practical capacity of a lithium–iodine pacemaker battery (Enertec Alpha 33) as a function of current drain. (By courtesy of Medtronic)

7.4 Lithium–lead iodide, lead sulphide batteries

In this primary system, the use of non-volatile salts as active cathode materials circumvents the problems of self-discharge, etc. associated with the iodine volatility of polyiodide or iodine charge transfer cathodes. In addition, development of a lithium ion solid electrolyte of greater conductance than LiI has permitted cells to be built in the form of compressed pellets without the creation of prohibitive values of internal resistance.

Mixtures of ion-conducting and non-conducting phases have been termed **polyphase solid electrolytes**. Mixing LiI with approximately 50 m/o γ-Al_2O_3 increases the conductivity of the lithium ions by a factor of about twenty, according to some workers, and of over a hundred, according to others. The origins of this enhancement in conductivity are not clear, and cannot be explained by any classical doping mechanism which would require an increase in the number of lithium ion vacancies in the lattice. Despite the complete stability of the polyphase system to lithium, the presence of small amounts of water in the alumina particles has been suggested as an important factor. It is considered that adsorbed water might react with the lithium iodide to form a highly conductive monohydrate phase. It is known that the conductance of LiI increases by about two orders of magnitude when passing from the anhydrous salt to $LiI.H_2O$. However it is unlikely that the contribution of trace levels of water alone can explain the enhancement in conductivity observed in $LiI(Al_2O_3)$ and it may be that a mechanism involving increased grain-boundary or surface conductivity, favoured by the high surface area of the dispersed alumina, should be considered.

The cell may be written as

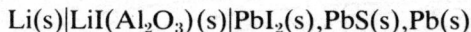

$$Li(s)|LiI(Al_2O_3)(s)|PbI_2(s),PbS(s),Pb(s)$$

and has now an OCV of 2.0 V at 25°C, dropping to 1.91 V after the first few percent of the discharge. The discharge reactions are

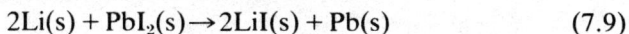

$$2Li(s) + PbI_2(s) \rightarrow 2LiI(s) + Pb(s) \tag{7.9}$$

and

$$2Li(s) + PbS(s) \rightarrow Li_2S(s) + Pb(s) \tag{7.10}$$

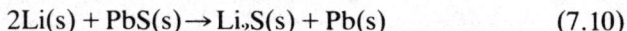

The cells have a virtually 100% discharge efficiency, and provided that they are enclosed in hermetically sealed cases, they are probably the most stable of all active primary batteries.

A schematic cut-away view of the practical cell developed by P.R. Mallory & Co. (now Duracell International) is shown in Fig. 7.18. The electrolyte layer is approximately 0.2 mm thick. The cell pellet is retained in a

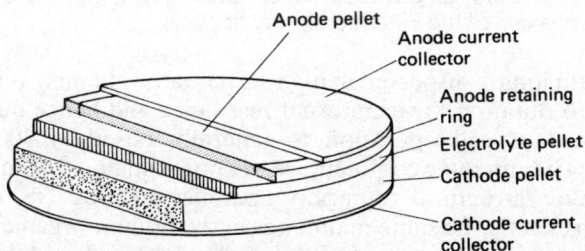

Anode pellet

Anode current collector

Anode retaining ring

Electrolyte pellet

Cathode pellet

Cathode current collector

7.18 Cross-section of a $Li(s)|LiI(Al_2O_3)(s)|PbI_2(s),Pb(s)$ solid-state cell. (By permission of the Electrochemical Society)

polypropylene ring and sealed in a steel case. Because of the critical effect of water vapour on both the anode and electrolyte, the cells are manufactured in an atmosphere where the water level is less than 15 ppm. The cell is useful for low rate applications at normal or slightly elevated temperatures; at high rates it becomes very inefficient. Because of its long term stability and reliability, this cell has found application in the cardiac pacemaker field. A typical pacemaker battery has three cells in series in the one battery case, giving an OCV of 6 V and a rated capacity of 140 mAh at a $2\mu A$ drain. The projected practical energy density is 490 Wh dm^{-3}. Batteries stored at 95°C for a year showed no change in discharge characteristics, so that self-discharge is considered to be negligible. Typical discharge curves at 37°C (body temperature) for this battery are shown in Fig. 7.19.

The use of As_2S_3 as an alternative cathode material has been proposed: this might give energy densities in the range 700–900 Wh dm^{-3}.

7.5 Lithium alloy–titanium disulphide and related batteries

One of the few examples of a practical rechargeable solid-state system also

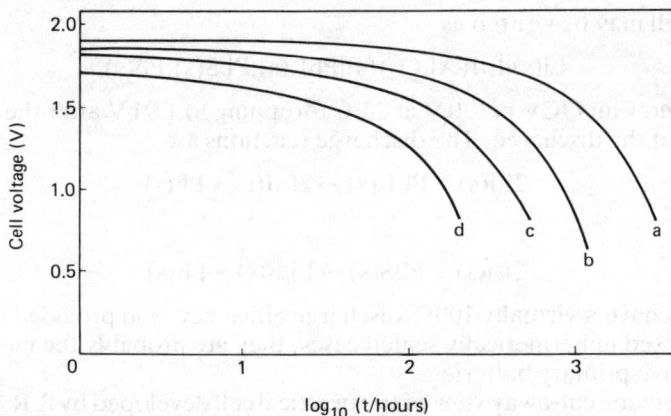

7.19 Discharge characteristics of 136 mAh lithium–lead sulphide, lead iodide solid-state cells under various loads: at 37 °C: (a) 18 μA; (b) 36 μA; (c) 54 μA; (d) 90 μA. (By permission of the Electrochemical Society)

uses lithium iodide dispersed in high surface area alumina as the electrolyte. In order to obtain a lower internal resistance and hence more power, the temperature of cell operation is generally raised to 300°C where the $LiI(Al_2O_3)$ has an ionic conductance of approximately $19\,Sm^{-1}$. Cells of this type employ 'structural retention' cathodes such as TiS_2 or TaS_2 which operate in exactly the same manner as in the lithium-organic cells described in Chapter 5. In this form of electrode, species such as lithium atoms can diffuse along the van der Waals' gaps in the layer structure of the electronically conducting host material, and as they do not accumulate at the interface, problems common to most metal/solid electrolyte interfaces, such as contact polarisation, dendritic growth, etc., are avoided. On the other hand, some limitations in rate are to be expected if diffusion within the electrode material is slow. At 300°C lithium metal is liquid, and it is therefore replaced by a lithium–silicon alloy which is solid, thus eliminating container problems and simplifying cell construction, and which has a reversible uptake/yield of lithium, as discussed in Chapter 6.

This battery system is being developed by P.R. Mallory and Co. (now Duracell International) who use a nominal $Li_{22}Si_5$ alloy and have investigated a number of cathode materials, including TiS_2, TaS_2 and a blended mixture of TiS_2 and Sb_2S_3. As illustrated schematically in Fig. 7.20 the cell is assembled by pressing the components into a single pellet, which is then hermetically sealed in a stainless steel case with glass-to-metal seal. The advantage of a 'soft' electrolyte such as $LiI(Al_2O_3)$ is that an effective cell can be made without the need for high temperature sintering processes, as good interparticle contact can be achieved simply by cold pressing (at 343 MPa). In addition, volume changes can be easily accommodated.

The effect of cathodic mixture composition is seen in the typical charge/discharge cycles shown in Fig. 7.21. Optimum behaviour was obtained with a mixture of the intercalation host with electrolyte and graphite which

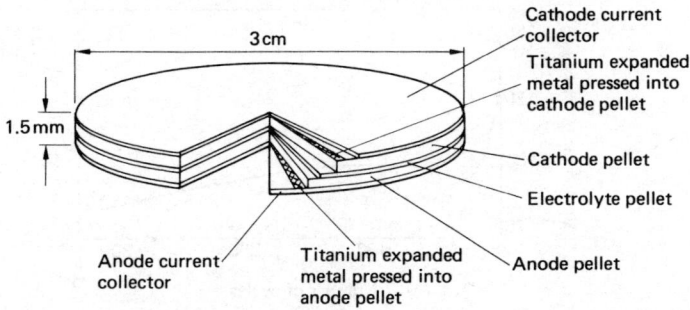

7.20 Cross-section of the $Li_nSi(s)|LiI(Al_2O_3)(s)|TiS_2(s),SbS_3(s),Bi(s)$ solid-state cell.

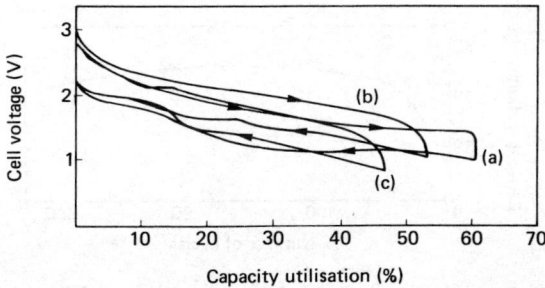

7.21 Effect of cathode composition on charge-discharge cycle at 300°C of a $Li_{4.4}Si|LiI(Al_2O_3)(s)|TiS_2(s),X(s)$ solid-state cell with 100 mAh capacity. Rate: C/10; cycle number: 15. Cathode composition (TiS_2,X) in weight per cent:
  (a) TiS_2,FeS, graphite, $LiI(Al_2O_3)$: 40/40/5/15
  (b) TiS_2, graphite, $LiI(Al_2O_3)$: 70/10/20
  (c) TiS_2,MoS_2, graphite, $LiI(Al_2O_3)$: 40/40/5/15
  (By courtesy of Duracell International)

respectively reduce contact polarisation and enhance the electronic conductance. A battery using this cathode constitution can be represented as

$$Li_nSi(s)|LiI(Al_2O_3)(s)|TiS_2(s),LiI(Al_2O_3)(s),C(s)$$

The e.m.f. of this cell is 2.4 V at 300°C and the cell reaction is given by

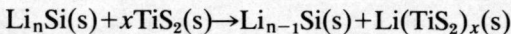

$$Li_nSi(s)+xTiS_2(s)\rightarrow Li_{n-1}Si(s)+Li(TiS_2)_x(s)$$

The effect of cycle life on coulombic efficiency and capacity utilisation is shown in Fig. 7.22 and 7.23. Over 300 cycles at current densities of $5\,mA\,cm^{-2}$ may be obtained, with efficiencies of 80–90% for a 400 mAh rated cell. The practical energy density is $280\,Wh\,kg^{-1}$.

The good cycling behaviour and energetic characteristics, coupled with the inherent advantages of an all solid-state system make this system a promising candidate for large scale storage applications. The total absence of a liquid phase makes individual cell containers unnecessary in a multicell battery configuration, thus leading to simplified structure and optimum

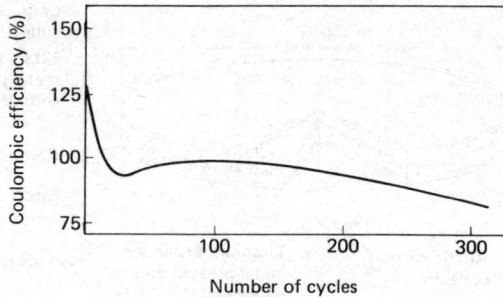

7.22 Effect of cycle life on coulombic efficiency for $Li_{4.4}Si/TiS_2$ solid-state cells. (By courtesy of Duracell International)

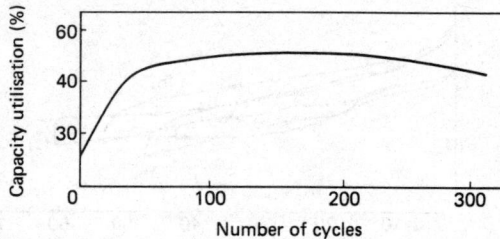

7.23 Effect of cycle life on capacity utilisation for $Li_{4.4}Si/TiS_2$ solid-state cells. (By courtesy of Duracell International)

packaging efficiency. Design studies have been made for a 72 kWh proto-type which uses twenty-four battery modules in parallel – each module contains thirty-six cells in series to give a 72 V system. The total energy density of the battery, including provision for heating elements and insulation, is estimated as 170 Wh kg^{-1} (310 Wh dm^{-3}). However additional data on durability and on prolonged cycling operation are required before a full evaluation of the system can be made and its competitiveness with longer established high temperature storage batteries (Chapter 6) can be assessed.

7.6 Hydrogen concentration cells

In the last few years a number of proton conducting solids have been discovered which show promise for application in electrolysis cells, fuel cells, sensors, electrochromic displays and secondary battery systems. Most of these electrolytes are unstable at high temperatures, but have con-ductances of up to 0.7 Sm^{-1} at 20°C. The best known of these materials, hydrogen uranyl phosphate tetrahydrate, $HUO_2PO_4.4H_2O$ – usually abbreviated as HUP – has been used in the development of a hydrogen concentration cell which may be used in an energy storage role. The cell is assembled by compressing a HUP disc between two metal hydride electrodes.

The cell based on the zirconium hydride–palladium hydride couple

$$Cu(s),ZrH_y(s)|HUP|PdH_x(s),Cu(s)$$

has an e.m.f. of 0.6 V at 25°C in its fully charged state. Polarisation effects are only moderate, and the 'soft' nature of the electrolyte accommodates differential expansion and contraction thus maintaining good interfacial contact on cycling. Capacity losses are however experienced on long term cycling.

7.7 Lead–cupric fluoride thin layer batteries

This is a system which is interesting mainly because of its conformation and its method of fabrication. The cell is based on the fluoride ion solid conductor, β-PbF_2. The ionic conductance of this material is so low at room temperature ($\approx 5 \times 10^{-5} Sm^{-1}$) that its use as an electrolyte in the form of a compressed pellet would not be feasible because of the prohibitively high internal resistance of the resulting cell. Instead, complete thin film solid-state cells have been developed. A typical thin film cell such as

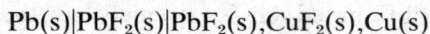

$$Pb(s)|PbF_2(s)|PbF_2(s),CuF_2(s),Cu(s)$$

is formed by evaporating thin layers of the anode material (Pb), the electrolyte (PbF_2), electrolyte-cathodic mixture (PbF_2–CuF_2) and the cathode current collector (Cu) sequentially onto a glass or quartz substrate. Here again the electrolyte and active electrode material are mixed to improve the ionic conductance of the cathodic mass, thus reducing interfacial polarisation. The net cell process is

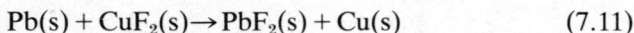

$$Pb(s) + CuF_2(s) \rightarrow PbF_2(s) + Cu(s) \tag{7.11}$$

with which is associated an e.m.f. of 0.70 V at 25°C.

Two different geometries have been used in the implementation of laboratory thin film batteries. As shown in Fig. 7.24 and 7.25, one form of battery is comprised of twelve small (0.1 cm²) cells and the other, of four larger (2.0 cm²) cells. Practical open circuit voltages range from 0.55 to 0.7 V, depending on the CuF_2 activity in the cathodic mass. (The co-deposition of CuF_2 and PbF_2 may give rise to a solid solution of the two components.)

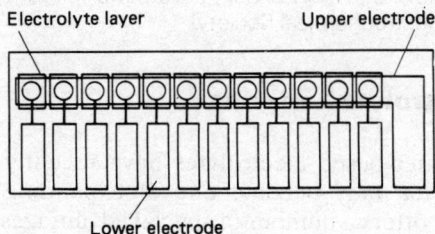

7.24 Thin film 0.1 cm² Pb/CuF₂ solid-state cells. (By permission of the Electrochemical Society)

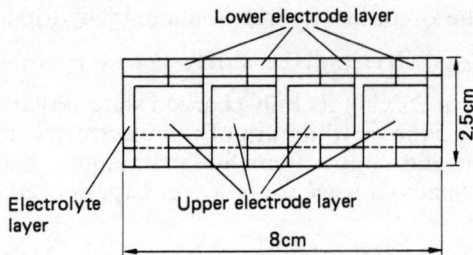

7.25 Thin film 2.0 cm² Pb/CuF₂ solid-state cells. (By permission of the Electrochemical Society)

Typical discharge curves of the $0.1\,cm^2$ cells through a $300\,k\Omega$ load are given in Fig. 7.26. It may be observed that the cells show a marked 'voltage delay' – i.e. a fall in voltage at the beginning of discharge, followed by a gradual recovery. This suggests an 'activation' process, usually associated with the dissolution of passive layers on the electrode surfaces. After the recovery, the voltage stabilises before decaying abruptly on the complete consumption of the (limiting) cathodic active material. While the power output from these PbF_2-based cells is extremely low, the concept of thin film cells is very important, since it points the way to extreme miniaturisation and to the possibility of self powered CMOS-type memories and other devices.

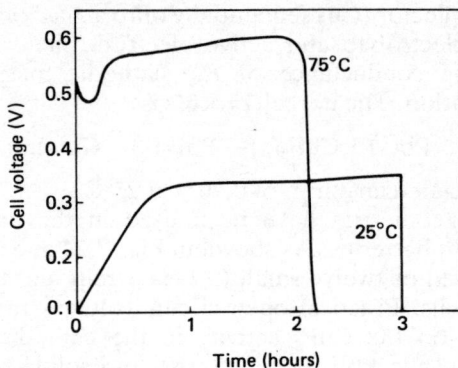

7.26 Discharge curves of thin film Pb/CuF₂ solid-state cells through 300 kΩ loads. (By permission of the Electrochemical Society)

7.8 Polymer electrolyte batteries

Batteries using polymer-based electrolytes have recently attracted much attention and this area may become one of expanding importance. A polymeric electrolyte offers a number of special advantages: the electrolytic phase can be readily formed with greatly reduced thickness, shaped in any desired configuration and is often basically of low cost. In addition, the mechanical flexibility of the polymer enables solid-state cells to be designed with optimised electrode/electrolyte interface configurations (for example, in 'composite' electrodes).

The most studied polymeric electrolytes up to the present time are complexes based on the combination of alkali metal salts with polyethers such as poly(ethylene oxide), PEO:

$$(-CH_2-CH_2-O-)_n$$

These complexes exhibit ionic conductivities in the range 10^{-5} to $10^{-3}\,Sm^{-1}$ at moderate temperatures (50–100°C). The conductance of a typical example, having an empirical formula of $(PEO)_5\,LiSCN$, is shown as a function of temperature in Fig. 7.27.

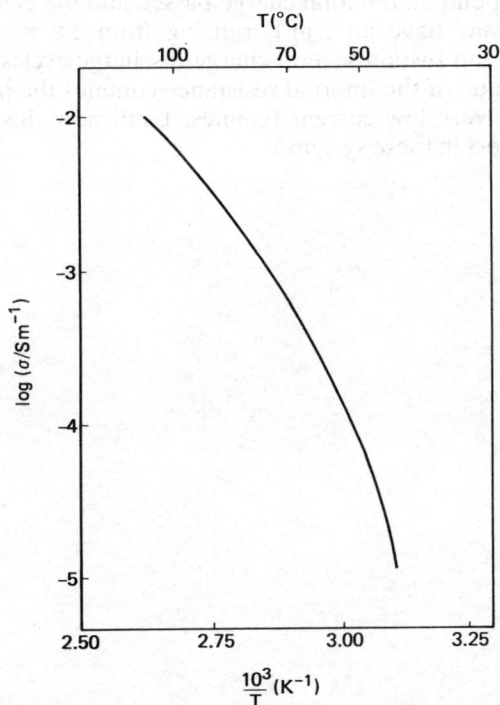

7.27 Conductivity of $(PEO)_5$ LiSCN polymeric electrolyte as a function of temperature.

Proposed structural models for semicrystalline Na^+ and Li^+ complexes suggest sequences of chains which form regular helical channels where the cations may reside. The simplest mechanism of conductance would then involve cation motion via vacancies along the helical channels. However 'inter-chain hopping' and fluctuations of the polymer chains also seem to play an important role in the conduction process.

A number of experimental battery types using polymeric electrolytes have been assembled and tested. A cell of the type

$$Li(s)|(PEO)_5LiC_2O_2F_3(s)|TiS_2(s)$$

has an e.m.f. of 2.6 V at 25°C, a value comparable with that observed for the Li/TiS$_2$ couple in organic electrolytes (see Section 5.5).

The possibility of using electrochemically doped electronically conducting poly(acetylene) as electrode material has resulted in the realization of 'all-polymeric' batteries. One example, in its fully discharged state is:

$$(CH)_x(s)|PEO.NaI(s)|(CH)_x(s)$$

Electrochemical doping of the poly(acetylene) occurs during the charging process, resulting in the cell

$$(CHNa_y)_x(s)|PEO.NaI(s)|(CHI_z)_x(s)$$

where y and z depend on the total charge passed and the electrode masses. Such a battery may have an e.m.f. ranging from 2.8 to 3.5 V at room temperature and can sustain several charge/discharge cycles. However the relatively high value of the internal resistance confines the performance of such batteries to very low current regimes. Until now this has restricted commercial interest in these systems.

8 Secondary hybrid cells

8.1 Introduction

A **hybrid cell** is defined here as a galvanic electrochemical generator in which one of the active reagents is in the gas phase. Hybrid cells occupy an intermediate position between the **closed-galvanic cells** described in the remainder of this book, where operation is confined to reactants added to the cell at manufacture, and **fuel cells** in which both cathodic and anodic reactants are supplied continuously (usually in gaseous form) from sources external to the cell. Hybrid cells take advantage of both battery and fuel cell technologies. In Section 3.7 the most common example of a primary hybrid cell, namely the metal–air system was considered, and a discussion of 'mechanically rechargeable' cells, where the spent metal anode is substituted by a new electrode at the end of the discharge, was included. This latter system is in reality a primary cell; in this Chapter, electrically rechargeable hybrid cells will be considered.

8.2 Metal–air cells

Introduction

Metal–air cells have a very favourable energy density which is achieved through not requiring to incorporate positive active components within the cell. To a large extent, the impetus for research and development in this field has arisen from possible EV applications where energy density is a critical parameter. Metals considered as possible secondary negative electrodes are confined to zinc and iron; others have been excluded on the grounds of cost or weight. The e.m.f. and theoretical energy density of zinc–oxygen and iron–oxygen couples are given in Table 8.1. To date there is still no commercially available secondary metal–air cell. However long term prospects for the zinc–air systems are considered good, since initial investment cost per kWh will be relatively low should high volume production become a reality.

The oxygen electrode

The principal features of the electrochemistry and construction of oxygen electrodes in alkaline solution considered in Section 3.7 for primary cells,

Table 8.1

E.m.f. and theoretical energy density of practical metal–oxygen cells

Couple	Reaction	E.m.f. V	Theoretical energy density Wh kg^{-1}
Zn/O$_2$	Zn(s) + ½O(g)→ZnO(s)	1.65	1090
Fe/O$_2$	Fe(s) + ½O$_2$(g)→FeO(s)	1.27	970

are common to secondary cells. If air is used rather than oxygen in the latter, it is then necessary to scrub the gas to remove carbon dioxide since otherwise the electrolyte becomes progressively contaminated with carbonate which reduces the conductivity and may block electrode pores. One manufacturer (Joseph Lucas Ltd.) developed a design incorporating a pumped air flow over the back of the porous electrode system which was able to maintain a stable gas-liquid interface without the use of water repellant surface treatment. Unfortunately the rates of water and heat removal from the cell were too high when this technique was used.

The main problem associated with oxygen electrode development for secondary batteries is related to the behaviour of the electrocatalyst during recharge, when it is necessary to apply high anodic voltages. Inexpensive catalysts based on carbon or silver cannot be used as they are rapidly oxidised. Even noble metal catalysts show significant dissolution or deep surface oxidation on recharge which reduces their catalytic efficiency for the discharge reaction and hence increases the oxygen electrode polarisation. One solution to this problem is the use of a third (auxiliary) electrode of perforated metal or wire mesh which carries all of the anodic current during charging. This procedure requires the mechanical switching of every cell in a battery twice a cycle, and is completely unsuitable for high power batteries since the size and cost of the switches would be prohibitive. An alternative method developed by Lucas, Siemens, Swedish National Development Co., and others is to use a composite electrode structure (Fig. 8.1 (a)–(b)). Here the auxiliary electrode is in parallel with their air electrode. During discharge the auxiliary is inactive, but during charging a pocket of oxygen bubbles is formed which effectively isolates the active oxygen electrode from the electrolyte. In Fig. 8.1 (a) the fine wire mesh on the solution side of the electrode fulfills the same function as the coarse pored structure in Fig. 8.1 (b).

The oxygen electrode suffers from considerable polarisation losses on discharge, largely due to mass transport limitations. Metal–air cells have consequently poor high rate performance. Polarisation is also high on charge (Fig. 8.2) so that cycle energy efficiency is very low. Further problems associated with this electrode system include water loss, especially at elevated temperatures, and leakage of electrolyte.

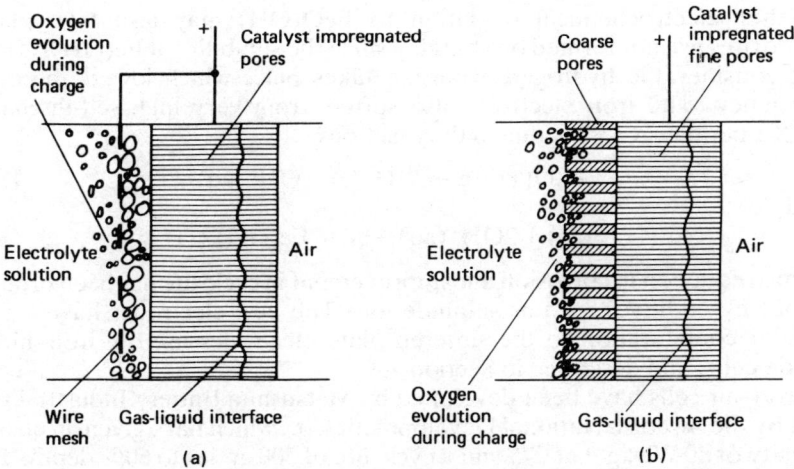

8.1 Methods for using layers of oxygen bubbles to isolate active oxygen electrode surfaces from electrolyte during charging.

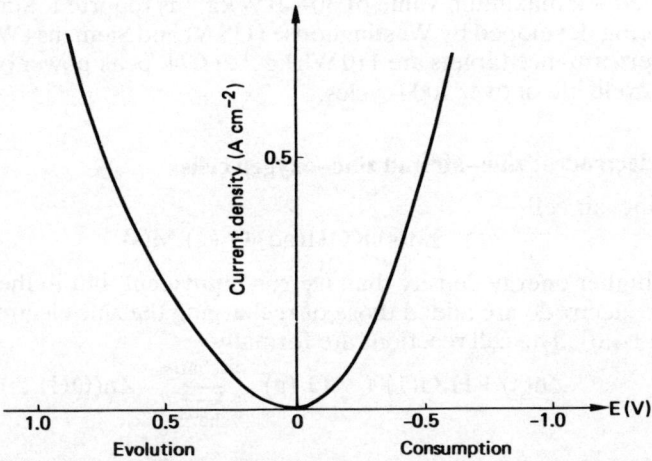

8.2 Anodic and cathodic overvoltage for oxygen evolution and consumption as a function of current density for a bright platinum electrode in oxygen saturated sulphuric acid.

Iron electrodes and iron–air cells

The iron–oxygen cells may be represented as

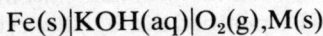

$$Fe(s)|KOH(aq)|O_2(g),M(s)$$

where M is the electrocatalyst, and the principal cell reactions are

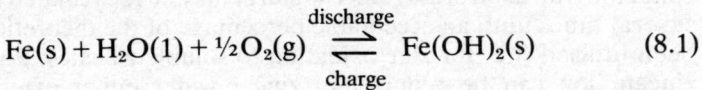

$$Fe(s) + H_2O(1) + \tfrac{1}{2}O_2(g) \underset{\text{charge}}{\overset{\text{discharge}}{\rightleftharpoons}} Fe(OH)_2(s) \qquad (8.1)$$

Further electrochemical oxidation to FeO(OH) may also take place. Dendrites are not formed on charging since the solubility of $Fe(OH)_2$ is low, but considerable hydrogen evolution takes place which lowers the cycle efficiency. The iron electrode also suffers from very high self-discharge ($\approx 2\%$ per day at 25°C), due to the reactions:

$$H_2O(1) + e \rightarrow \tfrac{1}{2} H_2(g) + OH^-(aq) \tag{8.2}$$

and

$$Fe(s) + 2OH^-(aq) - 2e \rightarrow Fe(OH)_2(s) \tag{8.3}$$

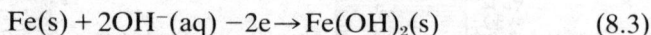

Some reduction in corrosion and improvement in cycle life has been brought about by additives such as sulphide ion. The best electrodes have a very similar configuration to the sintered plate electrodes used in iron–nickel oxide cells, and described in Section 4.4.

Iron–air cells have been developed by Matsushita Battery Industrial Co. and by the Swedish National Development Co., which have given an energy density of $80\,Wh\,kg^{-1}$ at C/5 and a cycle life of 200 cycles to 60% depth. The latter company have produced 15–30 kWh batteries for EV testing. One limitation of the iron–air system for this application is the low power density achieved – a maximum value of $30–40\,W\,kg^{-1}$ is reported. Similar cells are also being developed by Westinghouse (USA) and Siemens (W. Germany). The performance targets are $110\,Wh\,kg^{-1}$ at C/4, peak power of $100\,W\,kg^{-1}$, and a cycle life of over 1000 cycles.

Zinc electrodes, zinc–air and zinc–oxygen cells

The zinc–air cell,

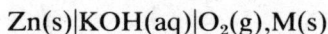

$$Zn(s)|KOH(aq)|O_2(g),M(s)$$

has a higher energy density than its iron equivalent, but to the problems of the air electrode are added those of recharging the zinc electrode (Sections 4.5 and 4.6). The cell reactions are formally

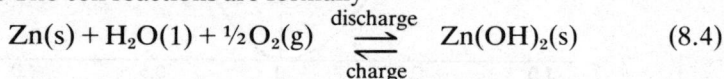

$$Zn(s) + H_2O(1) + \tfrac{1}{2}O_2(g) \underset{\text{charge}}{\overset{\text{discharge}}{\rightleftharpoons}} Zn(OH)_2(s) \tag{8.4}$$

but the main oxidation product of zinc is probably the $Zn(OH)_4^{2-}$ ion. A number of techniques have been tested with a view to improving the quality of the zinc deposit, such as electrolyte circulation, electrode vibration and the use of air bubblers to disturb the diffusion layer. Energy densities of over $130\,Wh\,kg^{-1}$ at C/5 have been claimed, but it is generally accepted that the zinc electrodes deteriorate fairly rapidly with cycling.

The most encouraging results have come from a system developed principally by CGE in France, and shown schematically in Fig. 8.3. A slurry of zinc powder in aqueous KOH is pumped through the cylindrical cell whose outer surface comprises the air electrode and for which the negative current collector is an axial brass rod. The slurry may be recirculated through the cell several times until an acceptable percentage of the theoretical capacity has been utilised (i.e. for zinc oxidation to soluble zincate). For recharge, the zincate ion can be reduced to zinc powder either externally or in an

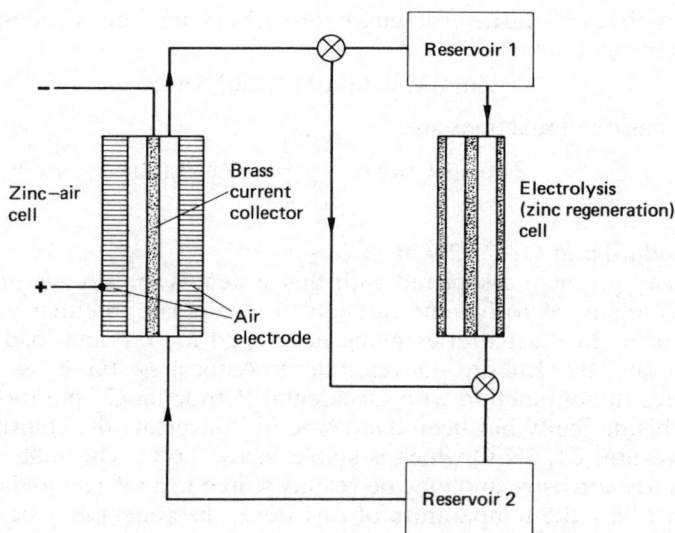

8.3 Schematic diagram of a zinc–air secondary battery system.

electrolysis cell connected to the system. In this context it is an advantage to deposit the zinc in dendritic form: the slurry is then reconstituted by momentarily reversing the current to detach the dendrites. The complete CGE system, including recharging cell and pumps is reported to have a practical energy density of over 90 Wh kg^{-1}. Cell lifetimes of over 1000 cycles have been obtained, with batteries of twelve series connected cells giving 500–600 cycles. However the complexity of the system together with the low specific power is likely to limit its exploitation as an EV power source.

An intriguing application suggested for a secondary zinc–oxygen battery is for energy storage on board spacecraft. In this elegant idea, the cells would be installed inside one of the oxygen tanks thus averting the need for gas supply pipes, valves, etc., and solving the problem of water loss. Cells would be recharged using solar converters.

8.3 Metal–halogen cells

Zinc–chlorine cells

The concept of an aqueous zinc–chlorine cell system brings together a number of attractive features. First, chlorine, as well as being abundant and inexpensive, has a highly positive equilibrium reduction potential so that cells with chlorine anodes have a high energy density. Further, chlorine is fairly soluble in aqueous solution (≈ 0.1 mol dm^{-3} at ambient temperatures, but dependent on the other solution components), so that three phase fuel cell electrode technology is not required. From the standpoint of the zinc electrode, zinc deposition morphology is much more favourable in acid media, with few dendrite problems unless at very high current densities. The cell is represented as:

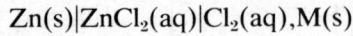

$$Zn(s)|ZnCl_2(aq)|Cl_2(aq),M(s)$$

and the basic cell reactions are:

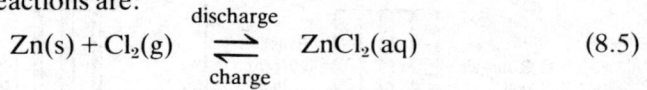

$$Zn(s) + Cl_2(g) \underset{charge}{\overset{discharge}{\rightleftharpoons}} ZnCl_2(aq) \tag{8.5}$$

The standard e.m.f. is 2.12 V at 25°C.

The key problem associated with this system is the storage of chlorine which is highly corrosive and difficult to manipulate in either gaseous or liquid form. In the batteries being developed for EV and load levelling applications by Energy Development Associates (Gulf & Western Industries in conjunction with Occidental Petroleum Corporation) in the USA, this difficulty has been overcome by converting the chlorine to the solid hydrate, $Cl_2.6H_2O$ which is stable below 9.6°C. This material is not particularly corrosive and may be readily stored in a refrigerated chamber. By controlling the temperature of this store, the system may be operated close to atmospheric pressure, without the need for corrosion resistant gas compressors.

A schematic diagram of the cell operating system is shown in Fig. 8.4. The

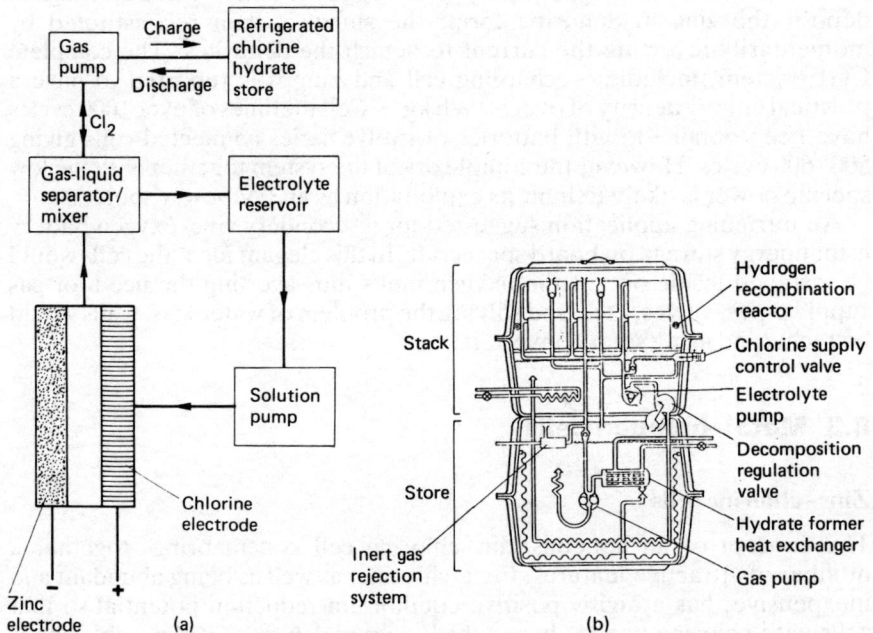

8.4 (a) Schematic diagram of a zinc–chlorine secondary battery system; (b) operational schematic of the BEST Facility battery module. (By permission of EPRI)

negative electrode is generally a solid graphite substrate on which zinc is deposited during charge, and the chlorine electrode is formed from either porous graphite or porous platinised or ruthenised titanium. There is no separator. The cell is assembled initially in an uncharged state and filled with an aqueous solution of zinc chloride at a concentration of approximately $4 \, mol \, dm^{-3}$ ($\approx 45\%$ by weight). As charging current is passed, zinc is deposited on the negative electrode and chlorine is evolved at the positive. Some coulombic losses occur due to chlorine migration and recombination with the zinc, but the majority of the chlorine produced is removed from the circulating electrolyte, converted to the solid hydrate by cooling and stored at 5–8°C. The electrolyte solution is replenished before returning to the cell and charging is terminated when the zinc chloride concentration has fallen to about $0.5 \, mol \, dm^{-3}$ ($\approx 15\%$ by weight). If the charged cell is now left on open circuit, chlorine dissolved in the electrolyte within the cell will react with zinc over an hour or so, reducing the OCV to zero. However as this amount of chlorine is relatively small, the degree of self-discharge is negligible. On discharge, chlorine is released from the hydrate by heating, using a heat exchanger and the heat released by the cell reaction and iR losses. The chlorine saturated electrolyte is pumped through the pores of the positive electrode where reduction to chloride ion takes place. In the discharge process coulombic losses are not so high as during charge, as chlorine tends to be removed from the electrolyte by the cell reaction before the flow reaches the zinc. It is normal to maintain the discharge until all the zinc has been oxidised: this results in a reproducible shape and morphology of the zinc deposit formed on the ensuing recharge.

The zinc electrode does not present the problems associated with charging in alkaline solutions. However the nature of the deposit is affected by factors such as current density, flow conditions, electrolyte concentration, etc., and various additives are used to minimise the incidence of dendrite formation. Zinc corrosion accompanied by hydrogen evolution is a significant problem because of the possibility of explosive hydrogen/chlorine mixtures being formed. Hydrogen evolution rates depend on the impurities present in the system and the pH of the electrolyte, which in turn is affected by zinc ion hydrolysis reactions and hence by $ZnCl_2$ concentration and temperature. Accumulation of hydrogen is avoided by regulating the electrolyte composition and by initiating controlled recombination using UV radiation.

Development of chlorine electrode materials has benefitted from the experience of chlor-alkali electrolysis cell technology. The main problem is to find the best compromise between cycle life and cost. Recently porous graphite, subjected to certain proprietary treatments, has been considered a preferable alternative to ruthenium-treated titanium substrates. The graphite electrode may undergo slow oxidative degradation, but this does not seem to be a significant process.

The practical batteries being developed by EDA are based on comb-type bipolar stacks, as shown in diagrammatic plan in Fig. 8.5(a). Machined graphite bus bars form the electrical connections between the electrodes on each stack and also separate the individual cell electrolytes. The zinc and chlorine electrode substrates are fitted into grooves on either side of the bus,

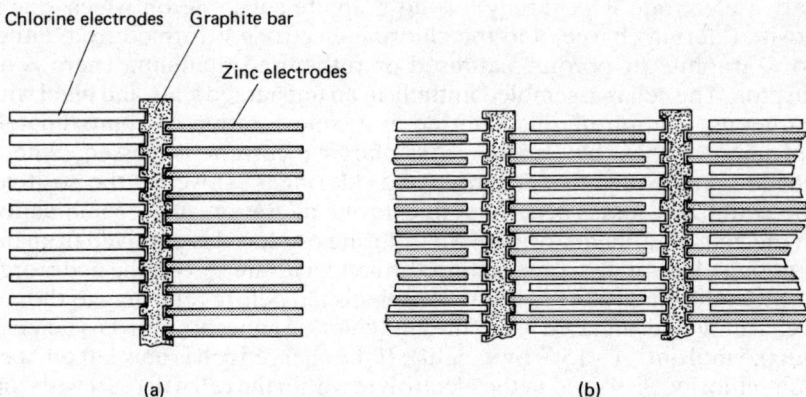

8.5   Comb-type bipolar electrodes for zinc–chlorine batteries: (a) bipolar stack; (b) unit cell.

and the battery is assembled by interleaving the comb structures as shown in Fig. 8.5(b). The stacks are terminated with monopolar units at each end. A channel is cut through the centre of the chlorine electrode substrate which enables electrolyte to be pumped through the porous mass during the charge or discharge process. The electrolyte then flows up between the zinc and chlorine electrodes and out at the end of each cell before returning to the pumping system. A great deal of work has been carried out in order to optimise the dimensions of the battery structure with a view to designing a system with a favourable current distribution and maximum capacity.

The coulombic efficiency of single cells is reported to be 70–80%. The principal loss, due to chlorine recombination with zinc, can be minimised by regulating the chlorine concentration and the rate of circulation of the electrolyte solution. Electrode polarisation is low both on discharge and charge (again provided that the electrolyte composition is controlled). The resulting flat charge/discharge curves are favourable for load levelling applications since simple converters can be used. Overdischarge does not create problems in zinc-limited cells since chlorine is evolved at the positive and consumed at the negative, so that a steady state condition develops. Overcharge is likely to result in zinc dendrites being formed and internally short-circuiting the cell. However since the amount of chlorine associated with the positive electrode at any instant is limited, there is no very significant self-discharge.

The complexity of the system and the dangers of chlorine evolution in the event of an accident disabling the refrigeration unit may militate against its application in EVs. However 10 MWh load levelling modules have been designed and are projected to have energy efficiencies of 70–75%, together with a very competitive capital cost per kW. Zinc–chlorine batteries are due to be evaluated at the BEST facility in the near future. Exact performance data are not easily found, but energy densities of 80–100 Wh kg^{-1} with peak power densities of up to 120 W kg^{-1} have been reported. More than one thousand deep cycles, have been demonstrated by 1.7 kWh batteries.

Zinc–bromine cells

The zinc–bromine system has a much lower energy density than that of zinc–chlorine, but the comparative ease of handling bromine and the absence of refrigeration requirements make it an attractive alternative. The relatively low power and energy density, coupled with the need for pumps, reservoirs, etc., probably rule out EV applications, but numerous development programmes for load levelling batteries are being actively pursued by a number of companies in the USA, in particular by Exxon Research and Engineering Corp., and by Gould, Inc.

The cell is represented as

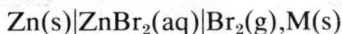

$$Zn(s)|ZnBr_2(aq)|Br_2(g),M(s)$$

and the overall cell reaction is written formally as

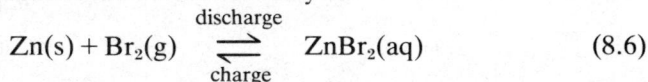

$$Zn(s) + Br_2(g) \underset{\text{charge}}{\overset{\text{discharge}}{\rightleftharpoons}} ZnBr_2(aq) \tag{8.6}$$

The standard e.m.f. is 1.83 V at 25°C and the theoretical energy density is 430 Wh kg^{-1}. In practice, complex species such as $Br_3{}^-$ are also involved, and the zinc ions also compete for the bromide ions to form a range of complexes from $ZnBr^+$ to $ZnBr_4{}^{2-}$. (At high zinc bromide concentrations the apparent cation transport number actually becomes negative). KCl may be added to the electrolyte to maintain high conductivity as the $ZnBr_2$ concentration falls during charge. The bromine electrode substrate material M, must maintain good reversibility for the electrochemical process, while at the same time resist chemical attack by the bromine. A number of materials such as titanium and ruthenium dioxide perform very well, but are expensive. Graphite cannot be used because of the formation of bromine intercalation compounds. Vitreous carbon resists attack and appears to be a very suitable material. While zinc can be deposited on a wide range of substrates, the choice in practice is very limited. As in the case of the zinc–chlorine cell, adequate control of zinc deposit morphology can only be retained by stripping off all the zinc from the positive electrode on all or most cycles. This requires that the cells are zinc–limited, and this in turn implies that in a battery, a number of cells must be reversed towards the end of a discharge to ensure complete stripping in the others. Under these circumstances, bromine is evolved on the negative, and the substrate must be able to withstand this process. In practical batteries this problem is often solved by using bipolar vitreous carbon electrodes.

A particular problem with this cell system is the high rate of self-discharge due to the chemical reaction of bromine with the zinc negative. One solution is to separate anolyte and catholyte by a cation-selective membrane which is impermeable to bromine, such as perfluorosulphonic acid-based materials like 'Nafion', as illustrated in Fig. 8.6(a). An alternative approach is to react the bromine in order to form an insoluble solid or liquid phase. For example, unsymmetric quaternary alkyl ammonium perchlorates react with bromine to form oily polybromides which significantly reduce the level of free bromine in solution. In some designs the insoluble bromine complexes are retained within the porous electrode structure; in more advanced large

Negative electrolyte reservoir — Positive electrolyte reservoir — Negative electrolyte reservoir — Positive electrolyte reservoir

Microporous separator

Bromine store

− + Cell stack

Membrane

− + Cell stack

Mixer

Negative loop   Positive loop   Negative loop   Positive loop

(a)   (b)

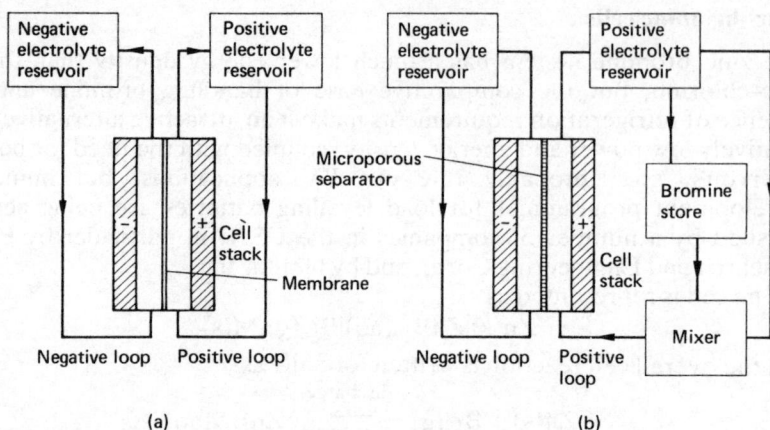

8.6 Schematic diagrams of zinc–bromine battery systems: (a) cell with cation selective membrane; (b) cell with reservoir for polybromide and microporous separator.

capacity systems, a separate polybromide reservoir is incorporated (Fig. 8.6(b)). A simple microporous separator is then sufficient to prevent excessive self-discharge. A typical cell stack is shown schematically in Fig. 8.7. Practical cells now all employ electrolyte circulation to improve the quality of the zinc deposit and to permit the operation of heat exchangers. Note that in both cell configurations shown in Fig. 8.6(a) and (b), twin circulation systems are required.

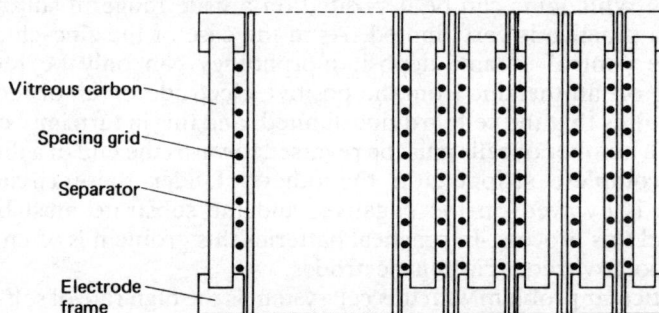

Vitreous carbon

Spacing grid

Separator

Electrode frame

8.7 Cell stack for a zinc–bromine battery.

Tests at the BEST facility of 5–10 MWh zinc–bromine modules are planned for 1985. At the present time, monopolar test cells have achieved over 2000 cycles. Batteries with energies of 20–80 kWh are currently being constructed with energy density predictions of 60–66 Wh kg^{-1}. Target values as high as 90 Wh kg^{-1} have been suggested.

8.4 Hydrogen–metal cells

As with the metal–oxygen system, hydrogen–metal cells can be considered as closed galvanic/fuel cell hybrids. They make use of the hydrogen electrode technology of fuel cells and that of nickel or silver electrodes from alkali secondary cells.

The charged hydrogen–nickel cell is represented as

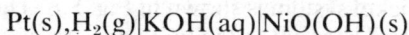

$$Pt(s), H_2(g) | KOH(aq) | NiO(OH)(s)$$

and has an OCV of 1.5–1.6 V in its fully charged state. The associated cell reactions are

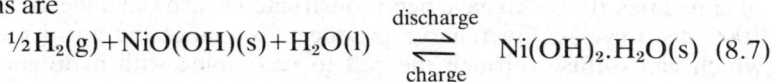

$$\tfrac{1}{2}H_2(g) + NiO(OH)(s) + H_2O(l) \underset{\text{charge}}{\overset{\text{discharge}}{\rightleftharpoons}} Ni(OH)_2 . H_2O(s) \quad (8.7)$$

The cell or battery is enclosed in a stainless steel or Inconel pressure vessel (Fig. 8.8). Hydrogen pressure rises from about 0.5 MPa in the fully discharged state to 3–10 MPa when charged, and the pressure in the vessel can be used to monitor the state of charge. Direct reaction between hydrogen and nickel oxide is relatively slow, but 6–12% of capacity is lost per day.

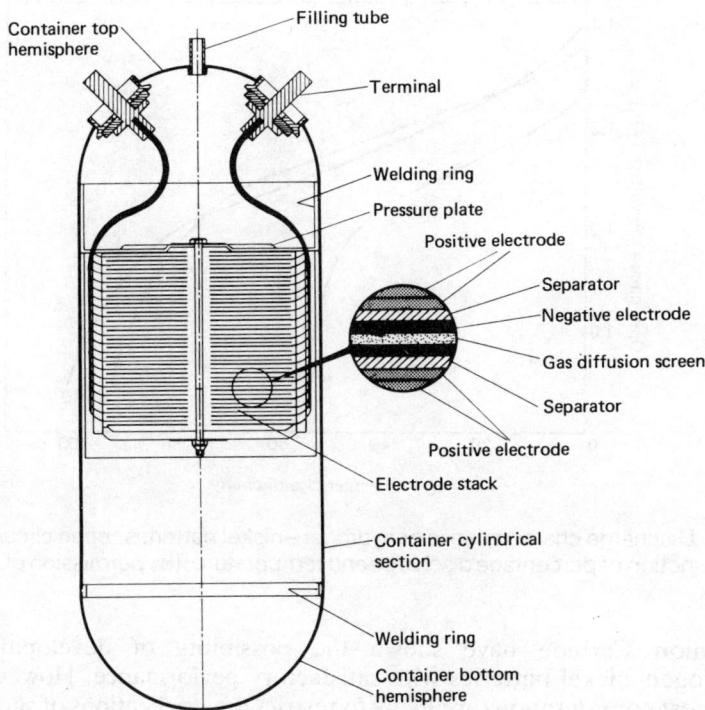

8.8 Cross-section of a SAFT hydrogen–nickel battery. (By permission of Academic Press)

The hydrogen electrode is generally a platinised or activated porous nickel mass which gives very good reversibility and low polarisation. The nickel oxide electrodes are chemically or electrochemically impregnated with nickel hydroxide, as in the sintered plate nickel–cadmium cells. The battery generally consists of a stack of nickel oxide and hydrogen electrodes separated by zirconium oxide cloth or asbestos felt separators used to immobilise the 30% KOH aqueous electrolyte. A schematic diagram of the SAFT hydrogen–nickel system is shown in Fig. 8.8. The hydrogen electrode substrate has a central channel to facilitate gas diffusion.

Characteristic charge and discharge curves are given in Fig. 8.9. Batteries of capacities 10–60 Ah have been constructed which can undergo more than 1000 deep cycles. Overcharge produces oxygen at the positive electrode which can diffuse through the cell to recombine with hydrogen to form water. Overdischarge causes hydrogen to be formed at the nickel oxide electrode at the same rate as it is being consumed at the hydrogen electrode, so that a steady state develops. Practical energy densities of this system are in the range 40–50 Wh kg^{-1}. The main application for batteries of this type is the replacement of nickel–cadmium cells in certain space applications, such as communications satellites, because of their higher energy density and longer life.

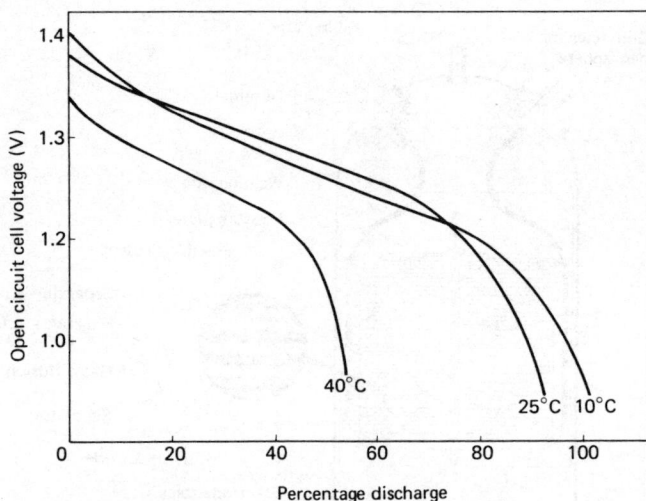

8.9 Discharge characteristics of hydrogen–nickel batteries: open circuit voltages as a function of percentage discharge and temperature. (By permission of Academic Press)

Union Carbide have shown the possibility of developing D-size hydrogen–nickel batteries with satisfactory performance. However safety and cost considerations are likely to restrict the applications of such units.

A new approach is to incorporate materials which can absorb large volumes of hydrogen. LaNi$_5$, for example can store up to seven mole percent of hydrogen. A relatively low pressure container can then be used with consequent reduction in weight and increase in energy density. Values of up to 65 Wh kg^{-1} have been claimed.

Hydrogen–silver cells are attractive (apart from cost) since the energy density is higher than the hydrogen–nickel analogue. The cell is

$$Pt(s), H_2(g)|KOH(aq)|AgO(s)$$

and the cell reactions are basically

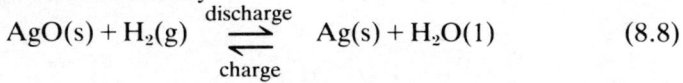

$$AgO(s) + H_2(g) \underset{\text{charge}}{\overset{\text{discharge}}{\rightleftharpoons}} Ag(s) + H_2O(l) \qquad (8.8)$$

The cell design is similar to that of the hydrogen–nickel system. There are problems with silver oxide solubility in the alkaline electrolyte but charge retention is comparable with that of nickel–cadmium.

Appendix 1

Operational modes and charging techniques for secondary batteries

The optimum method for recharging a battery depends both on the type and characteristics of the cell system under consideration, and on its service regime. The latter generally follows one of three basic patterns, determined by the application of the battery.

(a) Cyclic Operation

In applications such as load levelling or motive power, the battery is required to undergo a continuous sequence of deep discharges followed by recharge to maximum capacity. The prime requirement is therefore for a system which will provide a rapid and efficient charge, using as simple equipment as possible and avoiding damage to the battery. A number of methods may be considered.

(i) *Constant current.*

With this method, control equipment is simple and inexpensive and it is straightforward to calculate the charge delivered. However if high currents are used, problems are encountered with batteries which have porous electrode structures because of non-uniform current distribution and severe gassing may occur towards the end of the charge. At low current, the total charging time may be unacceptably long. This method can however be used with nickel–cadmium cells, which can sustain prolonged overcharging at $C/10$ without appreciable deterioration.

(ii) *Constant voltage.*

A constant charger is a statically controlled rectifier which uses a thyristor or similar device to correct for changes in output. The current is high at the beginning of the charge, and falls off as the charging proceeds, so that the current flow towards the end of the process is likely to be very low. If the charging voltage is too high, Joule heating early in the charge is likely to be excessive, leading to rapid deterioration of the cell.

Under certain circumstances, constant voltage charging of sealed secondary batteries can result in *thermal runaway*. In this phenomenon, at the end of the normal charge, overcharge cycle reactions (see Section 4.3) take over and the energy of the charging current is all converted to heat. The consequent rise in temperature of the cell lowers the overall polarisation and this increases the charging current, which in turn may raise the temperature further, and so on until an explosion results.

(iii) *Automatically regulated systems.*

There is now a wide range of regulated charging systems available which use techniques such as **step-wise reducing currents, constant voltage with current limitation, voltage taper charging**, combinations of constant current and constant voltage regimes, monitoring of OCV during current interruptions, etc. These procedures have been developed for individual cell systems to minimise charging time, while avoiding excessive side reactions towards the end of charge. One of the simplest methods is taper charging. Taper chargers have a constant high supply voltage and use a simple series resistor to modify the fall off in current as the battery voltage rises. An example of a sophisticated electronically controlled charger is the 'Spegel' system (Chloride Industrial Batteries) for lead–acid cells, which delivers a current that is inversely proportional to the cell voltage. The OCV is monitored every 15 minutes during a current interruption. If the voltage is below 2.3 V, the charge continues normally (Fig. A.1), but if it is higher, further application of charge is delayed. When the system recognises a fully charged condition, short current pulses are supplied to retain the battery in a fully charged state.

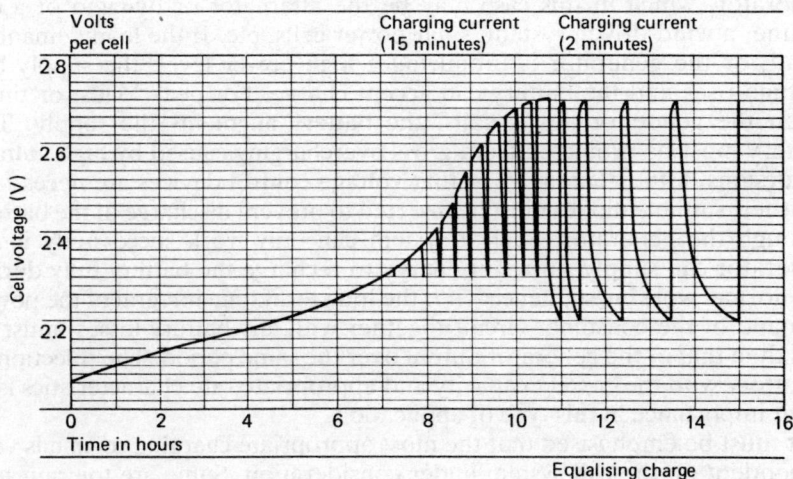

A.1 Typical recharge characteristics for a lead–acid cell under Spegel control. (By courtesy of Chloride Industrial Batteries.)

(b) Standby power

The consumer circuit is connected either in parallel to the mains unit and the battery, or the battery supplies power to the load by means of a relay which closes in the event of mains failure. In either case, the battery is only required to deliver current very occasionally during its entire working life. The normal charging requirement is therefore solely to compensate for self-discharge and thus to maintain the system at a state of full charge. This can be achieved by the technique known as **trickle charging:** the correct current input is one which exactly compensates for self-discharge and does not lead to gassing. Various empirical formulae are used for calculating the appropriate current level: for example, for lead–acid cells with Pb–Se grids a current of 1 mA per Ah of nominal capacity at C/10 is recommended for new cells, rising to 3 mA per Ah towards the end of cell life. It is necessary to incorporate in standby power installations circuits capable of high current (boost) charging following a deep emergency discharge. Standby batteries can also be **float charged** – i.e. battery, load and constant voltage charging circuit are connected in parallel and the latter can deliver sufficient output to meet the normal load demand plus any charging requirements. For lead–acid cells the voltage used is commonly 2.25 ± 0.05 V at ambient temperature. For very **low maintenance systems** (typically nickel–cadmium cells) where a long interval between electrolyte topping up is required, it is common to operate cells well below their fully charged state – e.g. nickel–cadmium cells may be float charged at 1.42 V.

(c) Continuous float or buffer mode operation

Again the consumer circuit is connected in parallel with the battery and a generator, which in this case may be the alternator or dynamo of a car engine, a wind-driven system, solar power cells, etc. If the load demand is small, or the generator is producing a high power level, the supply bus voltage rises and the battery can accept charge. For peak loads, or times when the generator power falls, the battery augments the supply. The battery must be protected from gross overcharging caused by high voltage excursions in the generator, so that voltage control devices are necessary. Further, suitable diodes must be inserted to prevent discharge of the battery through the generator. Such a system can only work successfully if the generator can supply sufficient energy to recharge the battery fully during the normal working sequence – i.e. the long term time integral of the power demand of the consumer circuit together with any battery losses, must be less than that of the generator output over the same period. The selection of a battery with the correct capacity and appropriate rate characteristics is of great importance in this type of application.

It must be emphasised that the most appropriate charging regime is very dependent on the cell system under consideration. Some are tolerant to a considerable amount of overcharging (e.g. nickel–cadmium batteries), while for others, such as zinc–silver oxide, overcharging can result in permanent damage to the cell. Sealed battery systems require special care: float charging should not be used and trickle charge rates should be strictly limited to the manufacturer's recommended values, since otherwise excessive cell temperatures or thermal runaway can result.

Appendix 2

Electrical quantities, physical constants and conversion factors

(i) Electrical quantities

Potential difference; e.m.f. : V (volt).
Electric field (gradient) : $V m^{-1}$ (volt per metre).
Resistance : Ω (ohm).
(Specific) resistance : Ωm (ohm metre).
Conductance : S (siemens); $1 S = 1 \Omega^{-1}$.
(Specific) conductance : $S m^{-1}$ (siemens per metre).
Electric charge : C (coulomb).
Current : A (ampere); $1 A = 1 C s^{-1}$.
Current density : $A m^{-2}$ (ampere per square metre).
Capacity : Ah (ampere hour); $1 Ah = 3600 C$.
    [N.B. The capacity of a battery (Section 2.5) measured in Ah, should not be confused with the *capacitance* of a capacitor (condenser) which is measured in F (farads).]
Capacity density or specific capacity : $Ah kg^{-1}$ (ampere hour per kilogram).
Volumetric specific capacity : $Ah dm^{-3}$ (ampere hour per cubic decimetre).
Power : W (watt); $1 W = 1 J s^{-1}$.
Power density or specific power : $W kg^{-1}$ (watt per kilogram).
Volumetric power density : $W dm^{-3}$ (watt per cubic decimetre).
Energy; work : J (joule); $1 Wh = 3.6 kJ$
                      $1 J = 1 C . V$
Energy density or specific energy : $Wh kg^{-1}$ (watt hours per kilogram); $1 Wh kg^{-1} = 3.6 kJ kg^{-1}$.
Volumetric energy density : $Wh dm^{-3}$ (watt hours per cubic decimetre); $1 Wh dm^{-3} = 3.6 kJ dm^{-3}$.

(ii) Physical constants

Avogadro number (L)	:	6.022×10^{23} molecules mol^{-1}.
Boltzmann's constant (k)	:	$1.381 \times 10^{-23} J K^{-1}$ molecule^{-1}.
Faraday's constant (F)	:	$9.649 \times 10^{4} C$ equiv^{-1}.
Planck's constant (h)	:	$6.622 \times 10^{-34} Js$.

Charge on electron (e)   : 1.601×10^{-19} C.
Gas constant (R)   : 8.314 J K^{-1} mol^{-1}.
2.303RT/F   : 5.916×10^{-2} V at 25°C.

(iii) Conversion factors

Length	1 inch	:	0.0254 m
	1 foot	:	0.3048 m
Area	1 sq. inch	:	6.45×10^{-4} m²
	1 sq. foot	:	0.0929 m²
Volume	1 cu. inch	:	1.639×10^{-5} m³
	1 cu. foot	:	0.0283 m³
	1 gallon (UK)	:	0.00455 m³
	1 gallon (US)	:	0.00379 m³
Mass	1 tonne	:	1000 kg
	1 lb	:	0.4536 kg
	1 ton (long)	:	1016 kg
Force	1 dyne	:	1.000×10^{-5} N
Pressure	1 atmosphere ⎱ 760 mm Hg ⎰	:	1.013×10^{5} Pa
	1 torr	:	1.333×10^{2} Pa
	1 bar	:	1.000×10^{5} Pa
Power	1 hp (horsepower)	:	746 W
Energy	1 eV	:	1.602×10^{-19} J
	1 erg	:	1.000×10^{-7} J
	1 cal	:	4.184 J
	1 Wh	:	3600 J
	1 BTU (British Thermal Unit)	:	1055 J (0.2931 Wh)
	1 mtoe (million ton oil equivalent)	:	4.187×10^{15} J (1.163×10^{13} Wh)
Charge	1 esu (statcoulomb)	:	3.333×10^{-10} C.

(iv) SI prefixes for submultiples and multiples

a	(atto)	10^{-18}		da	(deka)	10^{1}
f	(femto)	10^{-15}		h	(hecto)	10^{2}
p	(pico)	10^{-12}		k	(kilo)	10^{3}
n	(nano)	10^{-9}		M	(mega)	10^{6}
μ	(micro)	10^{-6}		G	(giga)	10^{9}
m	(milli)	10^{-3}		T	(tera)	10^{12}
c	(centi)	10^{-2}		P	(peta)	10^{15}
d	(deci)	10^{-1}		E	(exa)	10^{18}

Bibliography

General

G. Bianchi and T. Mussini, *Elettrochimica*, Tamburini Masson, Milan, 1976.
J.O'M. Bockris and A.K.N. Reddy, *Modern Electrochemistry*, Volumes 1 and 2, Macdonald Technical and Scientific, London, 1970.
J. Korita, J. Dvorak and V. Bohackova, *Electrochemistry*, Methuen, London, 1970.
G. Kortum, *Treatise on Electrochemistry*, Elsevier, Amsterdam, 1965.

W.M. Latimer, *Oxidation Potentials*, Prentice-Hall, N.J., 1952.
K.S. Pitzer and L. Brewer, *Thermodynamics*, (G.N. Lewis and M. Randall), McGraw-Hill, New York, 1961.
M. Pourbaix, *Atlas d'Equilibres Electro-Chimiques*, Gauthier-Villars, Paris, 1963.

W.J. Albery, *Electrode Kinetics*, Clarendon Press, Oxford, 1975.
H.H. Bauer, *Electrodics*, Thieme, Stuttgart, 1972.
P. Delahay, *Double Layer and Electrode Kinetics*, Wiley, New York, 1965.
H.R. Thirsk and J.A. Harrison, *Guide to the Study of Electrode Kinetics*, Academic Press, London, 1972.
K.J. Vetter, *Electrochemical Kinetics*, Academic Press, London, 1967.

R.N. Adams, *Electrochemistry at Solid Electrodes*, Dekker, New York, 1969.
J.P. Hoare, *Electrochemistry of Oxygen*, Wiley, New York, 1968.
A.T. Kuhn, *The Electrochemistry of Lead*, Academic Press, London, 1979.

D.H. Everett and F.S. Stone, *Structure and Properties of Porous Materials*, Butterworth Scientific, London, 1958.
S.J. Gregg and K.S.W. Sing, *Adsorption, Surface Area and Porosity*, Academic Press, London, 1967.

T. Erdey-Gruz, *Transport Phenomena in Aqueous Solutions*, Adam Hilger, London, 1974.
V.C. Levich, *Physicochemical Hydrodynamics*, Prentice Hall, N.J., 1962.

H. Harned and B. Owen, *The Physical Chemistry of Electrolytic Solutions*, Reingold, New York, 1958.

R.A. Robinson and R.H. Stokes, *Electrolyte Solutions*, Butterworth Scientific, London, 1959.

G. Charlot and B. Tremillon, *Chemical Reactions in Solvents and Melts*, Pergamon, Oxford, 1969.
A.K. Covington and T. Dickinson, *Physical Chemistry of Organic Solvent Systems*, Plenum Press, New York, 1973.
G.J. Janz and R.P.T. Tomkins, *Nonaqueous Electrolytes Handbook*, Volumes 1 and 2, Academic Press, London, 1972–3.
T.C. Waddington, *Nonaqueous Solvent Systems*, Academic Press, London, 1965.

Y.K. Delimarskii and B.F. Markov, *Electrochemistry of Fused Salts*, Sigma Press, Wolverhampton, 1961.
G.J. Janz, *Molten Salts Handbook*, Academic Press, London, 1967.
G. Mamantov, *Molten Salts: Characterisation and Analysis*, Dekker, New York, 1969.
G. Morand and J. Hladik, *Electrochimie des Sels Fondus*, Masson, Paris, 1969.

S. Geller, *Solid Electrolytes*, Springer, Berlin, 1977.
P. Hagenmuller and W. van Gool, *Solid Electrolytes*, Academic Press, London, 1978.
T. Takahashi and A. Kozawa, *Application of Solid Electrolytes*, J.E.C. Press, Cleveland, 1980.
W. van Gool, *Fast Ion Transport in Solids, Solid State Batteries and Devices*, North Holland, Amsterdam, 1973.
P. Vashishta, J.N. Mundy and G.K. Shenoy, *Fast Ion Transport in Solids: Electrodes and Electrolytes*, Elsevier North Holland, New York, 1979.

Batteries

P. Bauer, *Batteries for Space Power Systems*, NASA, Washington DC, 1968.
R.V. Bobker, *Zinc-in-Alkali Batteries*, The Society for Electrochemistry/ Joseph Lucas Ltd., University of Southampton, 1973.
H. Bode, *Lead–Acid Batteries*, Wiley, New York, 1977.
R.M. Dell, "Advanced Secondary Batteries: A Review", UK Atomic Energy Authority, Harwell, 1979.
S. Falk and A.J. Salkind, *Alkaline Storage Batteries*, Wiley, New York, 1969.
A. Fleisher and J.J. Lander, *Zinc–Silver Oxide Batteries*, Wiley, New York, 1971.
A.G. Garrett, *Batteries of Today*, Research Press, Dayton, Ohio, 1957.
D.P. Gregory, *Metal-air Batteries*, Mills and Boon, London, 1972.
G.W. Heise and N.C. Cahoon, *The Primary Battery, Volume 1*, Wiley, New York, 1971.
G.W. Heise and N.C. Cahoon, *The Primary Battery, Volume 2*, Wiley, New York, 1976.

R. Huber, *Trockenbatterien*, Varta Aktiengesellshaft, Hanover, 1972.
R. Jasinski, *High-energy Batteries*, Plenum Press, New York, 1969.
J. Jensen, P. McGeehin and R. Dell, *Electric Batteries for Energy Storage and Conservation*, Odense University Press, Odense, 1979.
K.V. Kordesch, *Batteries: Volume 1, Manganese Dioxide*, Dekker, New York, 1974.
K.V. Kordesh, *Batteries: Volume 2, Lead Acid Batteries and Electric Vehicles*, Dekker, New York, 1977.
L.F. Martin, *Dry Cell Batteries: Chemistry and Design*, Noyes Data Corp., Park Ridge, N.J., 1973.
D.W. Murphy, J. Broadhead and B.C.H. Steele, *Materials for Advanced Batteries*, Plenum Press, New York, 1980.
G.W. Vinal, *Primary Batteries*, Wiley, New York, 1950.
G.W. Vinal, *Storage Batteries*, McGraw-Hill, New York, 1955.
E. Witte, *Blei-und Stahlakkumulatoren*, Krausskopf-Verlag, Mainz, 1967.

International power sources symposia (Brighton)

D.H. Collins, Batteries, 1962, Pergamon Press, Oxford, 1963.
D.H. Collins, Batteries 2, 1964, Pergamon Press, Oxford, 1965.
D.H. Collins, Power Sources, 1966, Pergamon Press, Oxford, 1967.
D.H. Collins, Power Sources 2, 1968, Pergamon Press, Oxford, 1969.
D.H. Collins, Power Sources 3, 1970, Oriel Press, Newcastle-upon-Tyne, 1971.
D.H. Collins, Power Sources 4, 1972, Oriel Press, Newcastle-upon-Tyne, 1973.
D.H. Collins, Power Sources 5, 1974, Academic Press, London, 1975.
D.H. Collins, Power Sources 6, 1976, Academic Press, London, 1977.
J. Thompson, Power Sources 7, 1978, Academic Press, London, 1979.
J. Thompson, Power Sources 8, 1980, Academic Press, London, 1981.

Series

J. O'M. Brockris and B.E. Conway, *Modern Aspects of Electrochemistry*, Volumes 1–8, Butterworth Scientific, London, 1954–74. Volumes 9–13, Plenum Press, New York, 1975–79.
P. Delahay and C.W. Tobias, *Advances in Electrochemistry and Electrochemical Engineering*, Volumes 1–9, Wiley, New York, 1961–73.
H. Gerischer and C.W. Tobias, *Advances in Electrochemistry and Electrochemical Engineering*, Volumes 10–12, Wiley, New York, 1977–81.
A. Kozawa and K.V. Kordesch, *Progress in Batteries and Solar Cells*, Volumes 1–4, JEC Press, Cleveland, 1978–82.
E. Yeager and A.J. Salkind, *Techniques of Electrochemistry*, Volumes 1–3, Wiley, New York, 1973–8.

Reports, proceedings, etc.

Argonne National Laboratory Reports, 1975–80;
High Performance Batteries for EV Propulsion and Stationary Energy Storage

Symposium and Workshop on Advanced Research and Design
Lithium/Metal Sulphide Battery Development Programme

Commission of the European Communities: Report:-
Energy: Proceedings of the Meetings on Prospects for Battery Applications
and Subsequent R. and D. Requirements, Luxembourg, 1979.

DOE Report:-
Status of the DOE Battery and Electrochemical Technology Programme II,
R. Roberts, The Mitre Corporation, McLean, Va., 1980.

Electrochemical Society Proceedings:-
Electrode Materials and Processes for Energy Conversion and Storage,
1977. (Ed. J.D.E. McIntyre, S. Srinivason and F.G. Will).
Power Sources for Biomedical Implantable Applications and Ambient
Temperature Lithium Batteries, 1980. (Ed. B.B. Owens and N. Margalit).
Load Levelling, 1977. (Ed. N.P. Yao and J.R. Selman).
Energy Storage, 1976. (Ed. J.B. Berkowitz and H.P. Silverman).
Batteries Design and Optimisation, 1979. (Ed. S. Gross).

House of Lords Select Committee on Science and Technology: Report:-
Electric Vehicles, H.M. Stationary Office, London, 1980.

Lawrence Livermore Laboratory (UCRL – 52553):-
Energy Storage Systems for Automobile Propulsion, 1978. (E. Behrin).

Proceedings of Intersociety Energy Conversion Engineering Conferences,
Volumes 1–5, 1965–80.

Proceedings of the US Army Power Sources Symposium,
1970–80.

Specialist Periodical Reports:
Electrochemistry, The Chemical Society, London, Volumes 1–7, 1970–80.

Journals

Electrochimica Acta, International Society for Electrochemistry, Pergamon
Press, Oxford.
Electrokhimiya (English translation), Consultants Bureau, New York.
EPRI Journal.
Journal of Applied Electrochemistry, Chapman and Hall, London.
Journal of Electroanalytical Chemistry, Elsevier Sequoia S.A., Lausanne.
Journal of the Electrochemical Society, The Electrochemical Society,
Princeton, N.J.
Journal of Power Sources, Elsevier Sequoia, S.A., Lausanne.
Solid State Ionics, North-Holland, Amsterdam.

Glossary

activation overpotential (overvoltage): Contribution to the total overpotential due to the charge transfer step at the electrode interface.

active mass: The material in an electrochemical cell which takes part in the cell reaction. For example, the lead oxide contained in the positive plate of a lead–acid battery.

anode: The electrode at which oxidation takes place, and which gives up electrons to the external circuit.

anolyte: The electrolytic phase in contact with the anode.

battery: An assembly of two or more cells electrically connected to form a unit. For example, a 12 V SLI battery is made up of six 2 V cells in series. However, the term is also often used to indicate a single cell.

binder: A polymeric material added to the active mass to increase its mechanical strength.

bipolar electrode: An electrode assembly which functions as the anode of one cell on one side, and as the cathode of the next cell on the other side. Also known as a 'duplex' electrode, especially in Leclanché batteries.

bus, bus bar: A rigid metallic conductor which connects different elements of a battery; also, the conductor for an electrical system to which a battery terminal is attached.

button cell: Miniature cylindrical cell having a characteristic disc shape.

can, case: The external envelope of a cell or battery, or the box containing the cells and connectors.

capacity, rated: The value of the output capability of a battery, expressed in Ah, at a given discharge rate before the voltage falls below a given cut-off value, as indicated by the manufacturer.

cathode: The electrode at which reduction takes place and which withdraws electrons from the external circuit.

catholyte: The electrolytic phase in contact with the cathode.

cell: Electrochemical device which directly interconverts chemical and electrical energy.

cell reversal: Inversion of the polarity of the terminals of a cell in a multicell battery. Cell reversal is usually due to overdischarge, when differences in the capacity of individual cells result in one or more cells reaching complete discharge before the others.

charge acceptance: The ability of a secondary cell or battery to convert the active material to a dischargeable form. It is measured by the capacity which can be subsequently delivered to a load as a result of the charging process. If the charge acceptance is 100% then all of the electrical energy input would become available for useful output.

charge rate: See: C-rate.

charge retention: The ability of a charged cell to resist self-discharge.

charge, state of: The condition of a cell or a battery in terms of the remaining available capacity.

collector; current collector: Electronic conductor embedded in the active mass and connected to the bus bar or terminal.

concentration overpotential (overvoltage): Contribution to the total overpotential due to non-uniform concentrations in the electrolyte phase near the electrode surface caused by the passage of current.

corrosion: Oxidation of a metallic phase starting at the surface and caused by the reaction of the metal with components of the environment. In batteries corrosion phenomena play an important role especially in the case of primary aqueous cells and in high temperature systems.

C-rate: A method for expressing the rate of charge or discharge of a cell or battery. A cell discharging at a C-rate of τ will deliver its nominal rated capacity in $1/\tau$h; e.g. if the rated capacity is 2 Ah, a discharge rate of C/1 corresponds to a discharge current of 2A, a rate of C/10 to 0.2A, etc.

creep: The process by which liquid electrolytes, and in particular alkalies, can escape past rubber–metal or polymer–metal seals, or through minute cracks in a cell case or lid.

current density: Electric flux per unit area. It is generally defined in terms of the geometric or projected electrode area and is measured in $A\,m^{-2}$ or $mA\,cm^{-2}$.

cut-off voltage: Final voltage of a discharge or charge operation. In the case of discharge, it is chosen as the voltage value below which the connected equipment will not operate, or below which operation is not recommended because of the onset of irreversible processes in the cell. In the case of charge, it is selected to allow complete conversion of active material with a minimum of gassing.

cycle: Sequence of charge and discharge of a secondary battery.

cycle life: The total number of charge/discharge cycles that can be delivered by a secondary cell or battery, while maintaining a predetermined output capacity and cycle energy efficiency.

depolariser: A substance which is supposed to reduce electrode polarisation.

The term was introduced when it was believed that the electrode polarisation was due solely to gas evolution at the electrode and the action of the depolariser was to eliminate or prevent this process. Today views are different, but the term is still sometimes used, usually to denote the positive active material.

discharge curve: A plot of cell or battery voltage as a function of time, or of discharge capacity, under a defined discharge current or load.

discharge depth: The percentage of the capacity to which a cell or battery has been discharged. Shallow/deep discharge: small/large fraction of the usable capacity consumed.

discharge rate: See: C-rate.

drain: Withdrawal of current from a cell or battery.

dry cell: A cell in which the electrolyte is immobilised, being either in the form of a paste or gel or absorbed in a microporous separator material.

dry charged cell: A cell which is in its fully charged state but without electrolyte.

duplex electrode: Type of electrode system used in flat Leclanché multicell batteries, formed by zinc coated on one side with carbon. It acts as the cathode current collector for one cell and as the anode for the adjacent cell. (see 'bipolar electrode').

electrode: The electronic conductor and associated active materials at which an electrochemical reaction occurs.

electrodeposition: Deposition of a chemical species at the electrode of an electrolytic cell caused by the passage of electric current.

electrolysis: Chemical modifications (i.e. oxidation and reduction) produced by passing an electric current through an electrolyte.

electrolyte: The medium which permits ionic conduction between positive and negative electrodes of a cell. It may be solid or liquid. In some cases the electrolyte may take part in the cell reaction.

equalising charge: Passage of an amount of charge by which the undercharged cells of a battery are brought up to a fully charged condition without damaging those already fully charged.

expander: A substance added in small amount to the active materials of a lead–acid battery to improve the service life and capacity of the electrodes. In particular, an expander prevents the increase in crystal grain size of lead in the negative electrode.

failure: The state in which the performance of a cell or battery does not meet the normal specifications.

float charging: Method of recharging in which a secondary battery is continuously connected to a constant voltage supply that maintains the cell in fully charged condition.

forming; formation: A series of charge/discharge cycles carried out under carefully controlled conditions after the manufacture of a secondary cell in order to optimise the morphology of the active mass.

fuel cell: An electrochemical generator in which the reactants are stored externally and may be supplied continuously to the cell.

gassing: Gas evolution which takes place towards the end of the charging of a battery.

grain boundary: The surface separating two regions of a solid having different crystal orientations.

grid: The lead framework of a lead–acid battery plate which holds the active material in place.

group: A set of electrodes within a cell which are connected in parallel.

hybrid cell: Electrochemical cell in which **one** of the two active reagents is in the gas phase and may be supplied from an external source. A hybrid cell occupies an intermediate position between closed cells and fuel cells.

hybrid electric vehicle: A vehicle that has more than one type of power supply to support the drive: e.g. battery/motor plus fossil fuel/internal combustion engine.

immobilised electrolyte: See: dry cell.

inhibitor: A substance added to the electrolyte which prevents an electrochemical process, generally by modifying the surface state of an electrode. A well known example is that of corrosion inhibitors which prevent metal corrosion.

initial drain: Current that a cell or battery supplies when first placed on a fixed load.

internal resistance: Resistance to the flow of direct current within a cell, causing a drop in closed circuit voltage proportional to the current drain from the cell.

iR loss; iR drop: Decrease in the voltage of a cell during the passage of current due to the internal resistance of the bulk phases within the cell – mainly that of the electrolyte and the separators. Also known as 'ohmic loss'.

load: The external devices or circuit elements to which electric power is delivered by a cell or battery.

load levelling: The intervention aimed at reducing non-uniform conditions in electricity demand. The principle of load levelling is to store energy when demand is low and to use it to meet peak demand.

loss: See: *iR* loss; polarisation loss.

maintenance: The procedures which are required in order to keep a battery in proper operating conditions. They may include trickle-charging to compensate for self-discharge, addition of water to the electrolyte, etc.

mass transport: Transfer of materials consumed or formed in an electrode process to or from the electrode surface. Mechanisms of mass transport may include diffusion, convection and electromigration.

negative: Negatively charged electrode, usually of a secondary cell; acts as anode during discharge and cathode during charge.

ohmic loss: See: iR loss.

open circuit voltage: The voltage of a cell or battery under no-load condition, measured with a high impedance voltmeter or potentiometer.

overcharge: The continued application of charging current to a cell or battery after it has reached its maximum state of charge.

overdischarge: Forced discharge of a cell or a battery past 100% of the available capacity. In the case of a multicell battery, the overdischarge may cause **cell reversal**.

overpotential; overvoltage: Difference between the actual electrode voltage when a current is passing and the equilibrium (zero current) potential. A number of different effects may contribute to the total overvoltage.

oxidation: The loss of electrons by a chemical species.

passivation: Surface modifications of metallic materials which cause an increase in their resistance to corrosion process.

passivity: The condition of a metallic material corresponding to an immeasurably small rate of corrosion.

plate: In the terminology of secondary batteries, this has the same meaning as 'electrode'.

polarisation: Deviation from equilibrium conditions in an electrode or galvanic cell caused by the passage of current. It is related to the irreversible phenomena at the electrodes (electrode polarisation) or in the electrolytic phase (concentration polarisation).

polarisation loss: Reduction in the voltage of a cell delivering current from its equilibrium value.

positive: Positively charged electrode, usually of a secondary cell; acts as cathode during discharge and anode during charge.

post: See: terminal.

primary battery: A cell or battery whose useful life is over once its reactants have been consumed; i.e. one not designed to be recharged.

rate: See: C-rate.

recombining cell: A secondary cell in which provision has been made for the products of overcharge reactions to recombine so that no net change occurs to the composition of the cell system as a result of overcharging.

reduction: The gain of electrons by a chemical species.

reserve battery: In principle, any battery which will not deliver current in its manufactured form until activated by a suitable procedure, e.g. by adding the electrolyte to the dry components (water activated cells or cells activated by addition of special electrolytes), or by raising the temperature of the cell (thermal batteries, where the electrolyte is generally a mixture of salts in the solid state at ambient temperature).

reversal: See: cell reversal.

secondary battery; storage battery: Cell or battery which can be recharged after discharge, under specified conditions.

self discharge: Capacity loss of a cell or battery under open circuit conditions due to chemical reactions within the cell.

self discharge rate: The rate at which a cell or battery loses service capacity when standing idle.

separator: Electrically insulating layer of material which physically separates electrodes of opposite polarity. Separators must be permeable to the ions of the electrolyte and may also have the function of storing or immobilising the electrolyte.

service life: Timescale of satisfactory performance of a battery under a specified operating schedule, expressed in units of time or number of charge/discharge cycles.

shedding: The process whereby poorly adhering active mass (generally in the positive plate of a lead–acid cell) falls from the grid to form a sludge (mud) on the floor of the cell.

shelf-life: Period of time a cell can be kept idle after manufacture without significant deterioration.

short-circuit: The condition when the terminals of a cell or battery are connected directly.

SLI battery: A battery of usually 12 V or 24 V used for starting, lighting and ignition in vehicles with internal combustion engines.

stack: An assembly of parallel plates.

storage battery: See: secondary battery.

surface active agent: A substance which modifies the behaviour of a phase by interacting with its surface. For example, in the case of lead–acid batteries, the morphology of the active materials deposited at the electrodes may be strongly affected by the addition of surface-active agents.

terminal: The external electric connections of a cell or battery; also known as 'terminal post' or 'post'.

thermal battery: A type of reserve cell which is activated by raising the temperature.

thermal management: The means whereby a battery system is maintained within a specified temperature range while undergoing charge or discharge.

thermal runaway: A process in which a cell undergoes an uncontrolled rise in temperature due to the passage of increasing current (on, say, short-circuit discharge or constant voltage charging) as the temperature rises.

trickle charging: Method of recharging in which a secondary cell is either continuously or intermittently connected to a constant current supply in order to maintain the cell in fully or nearly fully charged condition.

uninterruptable power supply (UPS): A power system which maintains current flow without even a momentary break, in the event of mains or generator failure.

vent: Valve mechanism which allows controlled escape of gases generated during charging, but prevents spillage of electrolyte.

voltage delay: Time interval at the start of a discharge during which the working voltage of a cell is below its steady value. The phenomenon is generally due to the presence of passivating films on the negative electrode.

wet cell: A cell in which the liquid electrolyte is free-flowing.

Acknowledgements for figures reproduced from the literature

1.2	J.H. Tanne, *World Medicine*, 1981, *16* (25) 64.
3.29	P. Ruetschi, Plenary Lecture, ISE, Venice, 1980.
5.3	N. Margalit and H.J. Canning, Fall Meeting of the Electrochemical Society, Los Angeles, 1979; Proceedings of the Symposia on Power Sources for Biomedical Implantable Applications and Ambient Temperature Lithium Batteries, ed. B.B. Owens and N. Margalit, The Electrochemical Society, 1980, p. 339.
5.5	J. Bressan, A. De Guibert and G. Feuillade, ISE Meeting, Venice, 1980; Extended Abstracts, Volume 2, p. 737.
5.8	A. Morita, T. Iijima, T. Fujii and H. Ogawa, *J. Power Sources*, 1980, *5*, 111, Fig. 5.
5.18	A.J. Cuesta and D.D. Bump, Fall Meeting of the Electrochemical Society, Los Angeles, 1979; Proceedings of the Symposia on Power Sources for Biomedial Implantable Applications and Ambient Temperature Lithium Batteries, ed. B.B. Owens and N. Margalit, The Electrochemical Society, 1980, p. 95.
5.20	D. Linden and B. McDonald, *J. Power Sources*, 1980, *5*, 35, Fig. 4.
5.21	D. Linden and B. McDonald, *J. Power Sources*, 1980, *5*, 35, Fig. 5.
5.22	D. Linden and B. McDonald, *J. Power Sources*, 1980, *5*, 35, Fig. 7.
5.29	M.S. Whittingham, *J. Electrochem. Soc.*, 1976, *123*, 315.
5.30	M.S. Whittingham, *J. Electrochem. Soc.*, 1976, *123*, 315.
5.31	F.A. Trumbore, *Pure and Applied Chemistry*, 1979, *52*, 119, Fig. 2.
6.3	N.P. Yao, L.A. Heredy and R.C. Saunders, *J. Electrochem. Soc.*, 1971, *118*, 1039.
6.4–6.6	R.A. Sharma and R.N. Seefurth, *J. Electrochem. Soc.*, 1976, *123*, 1763.
6.8	A. Hooper, *J. Physics (D)*, 1977, *10*, 1493.
6.11–6.13	W.J. Walsh and H. Shimotake, *Power Sources 6*, ed. D.H. Collins, Academic Press, London, 1977.
6.17–6.18	N. Webber and J.T. Kummer, Proceedings of the 21st Annual Power Sources Conference, 1967.
6.23	F.G.R. Zobel, *J. Power Sources*, 1978, *3*, 29.
6.26	J.A. Asher, J.A. Bast and F.N. Mazandarany, Proceedings of the 15th Intersociety Energy Conversion Engineering

257

Conference, American Institute of Aeronautics and Astronautics, New York, 1980.

6.29–6.31 A.M. Chreitzberg, J.W. Consolly, M.R. Mamming and J.C. Skparchuck, *J. Power Sources*, 1978, *3*, 201.

6.32–6.33 G. Mamantov, R. Marassi, M. Matsunaga, Y. Ogata, J.P. Wiaux and E.J. Frazier, *J. Electrochem. Soc.*, 1980, *127*, 2319.

7.2 L.Y.Y. Chan and S. Geller, *J. Solid State Chem.*, 1977, *21*, 331.

7.5–7.7 B.B. Owens, *Advances in Electrochemistry and Electrochemical Engineering*, ed. P. Delahay and C.W. Tobias, Volume 8, John Wiley and Sons, New York, 1971, p. 1.

7.16–7.17 C.C. Liang and L.H. Barnette, *J. Electrochem. Soc.*, 1976, *123*, 453.

7.22–7.24 J.H. Kennedy and J.C. Hunter, *J. Electrochem. Soc.*, 1976, *123*, 10.

8.8–8.9 S. Font and G. Goulard, *Power Sources 5*, ed. D.H. Collins, Academic Press, London, 1975, p. 334.

Index